The Elephant
in the Classroom

The Elephant in the Classroom

Helping Children Learn and Love Maths

Jo Boaler

Souvenir Press

First published in the USA by Viking Penguin,
a member of the Penguin Group (USA) Inc under the title
What's Math Got To Do With It?

This first UK edition, under the title *The Elephant in the Classroom*,
has been substantially revised from the US edition.

First published in Great Britain in 2009 by Souvenir Press Ltd
43 Great Russell Street, London WC1B 3PD

This paperback edition 2010
Reprinted 2010 (three times)

ISBN 9780285638754

Typeset by M Rules
Printed in the UK by the MPG Books Group

For Colin Haysman

Acknowledgements

This book is the result of an incredible mathematical journey and the important relationships with colleagues, friends, teachers, and students that I enjoyed along the way. The journey started at King's College, London University, it continued at Stanford University in the USA, where I spent nearly ten years, and now continues at the University of Sussex. In all of these places I have learned from amazing students and teachers and these teachers and students are represented in the pages of this book.

This book was conceived during a year at a very special place, devoted to the generation of ideas: The Center for Advanced Study in the Behavioral Sciences, in California, where I was on sabbatical. I had given a presentation to the other fellows, a group of scholars who worked in different areas of social science research, on the results of my studies of mathematics learning. As the group heard about the effects of different forms of maths teaching, they responded strongly, with expressions of shock and dismay, and they urged me to get my results out to the general public. They encouraged me to write a book for the general public and many people – in particular Susan Shirk, Sam Popkin and David Clark – supported me along the way.

From that point I was greatly encouraged by my agent Jill Marsal, from the Sandra Dijkstra Literary Agency, and by Kathryn Court, my editor at Penguin USA who published the US version of this book. I wrote much of this book in the stimulating environment of Stanford's education school, surrounded by a group of graduate students who served as critics and supporters. I would personally like to thank all of my students, past and present, who contributed to the mathematics education group at Stanford. They are: Nikki Cleare, Jennifer DiBrienza, Jack Dieckmann, Nick Fiori, Melissa Gresalfi, Vicki Hand, Tesha Sengupta-Irving, Emily Shahan, Megan Staples, Megan Taylor, and Tobin White. Nick Fiori was my right hand person throughout the book, helping me with research, data collection, writing, mathematical thinking and editing. Nick is an exciting person to work with, he has a deep appreciation of the elegance and beauty in mathematics, and I was fortunate to work closely with Nick on this book.

This edition of the book was made possible by the vision of an incredible man – Ernest Hecht, who read my US book and has worked to bring the research and ideas to a broader audience. I have really appreciated Ernest's careful attention to all of the details in the book and the ways in which they may be communicated most effectively. It has been a pleasure to work with all of the team at Souvenir Press.

I have learned a great deal from some truly inspirational mathematics teachers in recent years – among them Cathy Humphreys, Carlos Cabanas, Estelle Woodbury, and Ruth Parker. They change students' lives on a daily basis, and I am fortunate to have been able to work with them and learn from them. I am also deeply grateful to the students of Railside, Greendale, Hilltop, Amber Hill and Phoenix Park schools, they all gave me their honest and insightful feedback on their mathematics teaching experiences and they have paved the way for a much better mathematical future for thousands of students. They are the reason that I wrote this book.

I have learned from some great teachers in my life – including Professors Paul Black and Dylan Wiliam, both of whom encouraged me greatly at an early point in my academic career and kindly read chapters of this book for me. Leone Burton, one of my strongest supporters, died recently; she will be greatly missed, by me and many others in mathematics education who appreciated her sharp, no-nonsense approach and her support of women. I am also indebted to many of my colleagues at Stanford University, in particular Professors Jim Greeno, Pam Grossman, Aki Murata, Rich Shavelson, and Deborah Stipek, for their unwavering support and collegiality.

Colin is always my rock – he looks after me, feeds me, puts up with me, and inspires me. He has great ideas about education and I trust his judgement more than anyone. This book is dedicated to Colin, as well as the smaller members of our team – Jaime and Ariane.

Contents

The Elephant
in the Classroom

Introduction.

Understanding the Urgency.

Far too many students hate maths. As a result adults all over the world fear maths and avoid it at all costs. Mathematics plays a unique role in the learning of most children – it is the subject that can make them feel both helpless and stupid. Maths, more than any other subject, has the power to crush children's confidence, and to deter them from learning important methods and tools for many years to come. But things could be completely different and maths could be a source of great pleasure and confidence for people. There is a huge gap between what we know works for children and what happens in most classrooms. We have the knowledge of good teaching methods, for schools and the home, but this essential knowledge is usually shared between experts in universities and communicated in scientific research journals. It is time that this important knowledge is communicated to a broader public – to parents who want to help their children and to teachers and other educators. This book will provide that knowledge, through real-life stories of children learning maths well, and details of the real maths that children should be learning. If you have the knowledge of good teaching methods and important maths learning principles, you can be very powerful in helping children, or yourself, to have a much brighter mathematical future.

There is an elephant in the room is an expression that refers to an idea that is very important but not talked about. I have called this book "The elephant in the classroom" because there is often a very large elephant standing in the corner of maths classrooms. The elephant, or the common idea that is extremely harmful to children, is the belief that success in maths is a sign of general intelligence and that some people can do maths and some people can't. Even maths teachers (the not so good ones) often think that their job is to sort out those who can do maths, from those who can't. This idea is completely wrong and this is why. In many maths classrooms a very narrow subject is taught to children, that is nothing like the maths of the world or the maths that mathematicians use. This narrow subject involves copying methods that teachers demonstrate and reproducing them accurately, over and over again. Of course very few people are good at working in such a narrow way, and usually everyone knows which people are good at it and which people are not. But this narrow subject is not mathematics, it is a strange mutated version of the subject that is taught in schools. When the real mathematics is taught instead – the whole subject that involves problem solving, creating ideas and representations, exploring puzzles, discussing methods and many different ways of working, then many more people are successful. This is the classic "win-win" situation – teaching real mathematics, means teaching the authentic version of the subject and giving children a taste of high level mathematical work, it also means that many more children will be successful in school and in life.

Although this seems to make perfect sense, (why wouldn't anyone want to rid schools of fake maths and bring in the real maths that children enjoy?) changing maths teaching in schools has proved to be very difficult and many children are still subjected to an out-dated and narrow form of teaching.

A few years ago when I was a professor at Stanford University, in California, I watched a "math" lesson (as they call it in the US) that I shall never forget. Several people had recommended that I visit the class to see an amazing and unusual teacher, and as I climbed the steps of the portable classroom I was curious about what I would find inside. I knocked on the door. Nobody answered so I pushed the door open and stepped inside. Emily's class was not as quiet as most maths classes I have visited, in America or England. A group of tall adolescent boys stood at the front, smiling and laughing, as they worked on a maths problem. One of the boys spoke excitedly as he jumped around explaining his ideas. Sunlight streamed through the windows, giving the front of the room a stage-like quality. I moved quietly among the rows of students to take a seat at the side of the room.

Emily Moskam, the teacher, nodded briefly in my direction in recognition of my arrival. All eyes were on the front, and I realized that the students had not heard the door because they were deeply engaged in a problem Emily had sketched on the board. They were working out the time it would take a skateboarder to crash into a padded wall after holding onto a spinning merry-go-round and letting go at a certain point. The problem was complicated, involving high-level mathematics. Nobody had the solution, but various students were offering ideas. After the boys sat down, three girls went to the board and added to the boys' work, taking their ideas further. Ryan, a tall boy was sitting at the back and he asked them: "What was your goal for the end product?" The three girls explained that they were first finding the rate that the skateboarder was traveling.

After that, they would try to find the distance from the merry-go-round to the wall. From there things moved quickly and animatedly in the class. Different students went to the board, sometimes in pairs or groups, sometimes alone, to share their ideas. Within ten minutes, the class had solved the problem drawing from trigonometry and geometry, similar triangles, and tangent lines. The students had worked together like a well-oiled machine, connecting different mathematical ideas as they worked towards a solution. The maths was hard and I was impressed. (*The full question and the solution for this maths problem, as well as the other puzzles in this book, are in the appendix*).

Unusually for a maths class it was the students, not the teacher, who had solved the problem. Most students in the class had contributed something, and they had been excited about their work. As the students shared ideas, others listened carefully and built upon them.

There have been many debates over the years between those who believe that mathematics should be taught traditionally, with the teacher explaining methods and the students watching and then practicing them, in silence, and those who believe that students should be more involved – discussing ideas and solving complex problems. Those in the 'traditional' camp have worried that student-centered teaching approaches would sacrifice standard methods, mathematical correctness, or high-level work. But this class was a perfect example of one that would please people on both sides of the debate, as the students fluently made use of high-level mathematics, which they applied with precision. At the same time, the students were actively involved in their learning and were able to offer their own thoughts in solving problems. This class worked so well because students were given problems that interested and challenged them and they were allowed to spend part of each lesson working alone and part of each lesson talking with each other and sharing ideas. As the students filed out of the room at the end of class, one of the boys sighed, "I love this class." His friend agreed.

Unfortunately, very few maths classes are like Emily Moskam's, and their scarcity is part of the problem with maths education. Instead of classes where students are actively engaged in mathematical problem solving, most school children watch a teacher demonstrate methods that they neither understand nor care about. Far too many students *hate* maths and for many it is a source of anxiety and fear. This is particularly true for girls. In England, only 38% of the students who choose A-level maths are girls[1], which is a shocking statistic, especially when we consider other countries such as the US where half of high-level maths students are women, even at university level. The Think Tank *Reform* recently claimed that over three-quarters of 16 year olds leave school in England without even an *elementary* grasp of maths[2]. More worryingly perhaps, students in schools are often made to feel inadequate in maths, from a very young age, which results in their developing very negative views of the subject. Something is going badly wrong with maths in our country and it is important that we do something about it, for the sake of our society and our children's future. Consider these chilling facts:

- Between 1989 and 2007 the proportion of students choosing A-level maths fell by two-thirds[3]
- An estimated 15 million adults in England struggle with basic mathematics[4]
- When UNICEF assessed students' sense of subjective well being, drawing on children's feelings about school, the UK came in last place, out of 20 countries – this has a lot to do with children's maths treatment in schools[5]
- Maths is widely hated by schoolchildren, surveys of school children report that it is the second most boring subject in school (behind science)[6].

- In 2000, in an international test of maths the UK was placed eighth – in 2007 the UK had dropped to 24th place[7].

Achievement and interest among children is low, but the problem does not stop there. Mathematics is widely hated among adults because of their school experiences, and most adults avoid maths at all costs. Yet the advent of new technologies means that we all need to be able to reason mathematically in order to work and live in today's society. What's more maths could be a source of great pleasure for people, if only adults could forget past experiences and see maths for what it is rather than the distorted image that was presented in school. When I tell people that I am a professor of maths education, they often shriek in horror, saying that they cannot do maths to save their lives. This always makes me sad because I know that they must have experienced bad maths teaching. I recently interviewed a group of young adults in England who had hated maths in school and were surprised at how interesting maths was in their work; many of them had even started working on maths puzzles in their spare time. They could not understand why their school had misrepresented the subject so badly.

This aversion to maths is reflected in our popular culture as well. Every time a child is shown struggling with homework on TV, sighing, slumped over crumpled exercise pages, we know the homework is going to be maths. When Bart Simpson, from *The Simpsons* returned his school textbooks to his teacher at the end of the school year, he proudly told his teacher that his maths book was "still in its original wrapping". When a speaking Barbie was first introduced to the world, some of her first words were "Math class is tough". The manufacturers quickly and correctly withdrew this feature following outcries from educators and parents. But Barbie and Bart are not alone. In England, maths is just as widely hated by school children. When Bliss magazine surveyed the views of schoolchildren they found that maths teachers were voted the most 'evil', presumably because of the hours of boring lessons that they make students endure[8].

Then again, amid this picture of widespread disdain there is evidence that maths has the potential to be quite appealing. Recent films depicting maths and mathematicians – such as 21, A Beautiful Mind, Good Will Hunting, and Proof – were all box-office hits and the mathematical TV show NUMB3RS consistently draws large audiences. Books such as Fermat's Enigma and Pi were best-sellers and Sodoku, the ancient Japanese number puzzle, has gripped people in the UK. Adults everywhere can be seen hunched over their number grids before, during and after work, engaging in the most mathematical of acts – logical thinking. These trends suggest something interesting: school maths is widely hated, but the mathematics of life, work and leisure is intriguing and much more enjoyable. There are two versions of maths in the lives of many people: the strange and boring subject that they encountered in classrooms and an interesting set of ideas that is the maths of the world, and is curiously different and surprisingly engaging. The purpose of this book is to introduce this second version to today's students, get them excited about maths, and prepare them for the future.

The Mathematics of Work and Life.

It is estimated that there will be twenty million more jobs in the future for people who are mathematical problem solvers. Unfortunately it has also been estimated that sixty percent of all new jobs in the twenty first century will require skills that are possessed by only twenty percent of the current workforce[9]. Employment surveys report that 50% of employers are dissatisfied with the basic numeracy of English school leavers[10]. But what sort of mathematics will young people need in the future?

Ray Peacock, research director of Phillips Laboratories in the UK, reflected upon the qualities needed in the high tech workplace:

> 'Lots of people think knowledge is what we want, and I don't believe that, because knowledge is astonishingly transitory. We don't employ people as knowledge bases, we employ people to actually do things or solve things . . . knowledge bases come out of books. So I want flexibility and continuous learning. (. . .) and I need team working. And part of team working is communications. (. . .)When you are out doing any job, in any business . . . the tasks are not 45 minutes max, they're usually 3 week dollops or one day dollops, or something, and the guy who gives up, oh sod it, you don't want him. So the things therefore are the flexibility, the team working, communications and the sheer persistence'[11]

Dr Peacock is not alone in valuing these employee capabilities. Surveys of employers from manufacturing, information technology and the skilled trades tell us that employers want young people who can use 'statistics and three-dimensional geometry, systems thinking and estimation skills. Even more importantly, they need the disposition to think through problems that blend quantitative work with verbal, visual and mechanical information; the capacity to interpret and present technical information; and the ability to deal with situations when something goes wrong.'[12]

Mathematical know-how is not only one of the most important qualities for workers to possess in the future, it is critical to successful functioning *in life*. As US researchers Forman and Steen say: 'today's news is not only grounded in quantitative issues (eg budgets, profits, inflation, global warming, weather probabilities) but it is also grounded in mathematical language (eg graphs, percentages, charts)'[13]. Whether browsing the web, interpreting medical records, administering medicine to children, reading the news, working with finances, or taking part in elections, 21st century citizens need mathematics. But the mathematics that people need is not the sort of mathematics learned in most classrooms – people do not need to regurgitate hundreds of standard methods, they need to reason and problem solve, flexibly applying methods in new situations. Mathematics is now so critical to young people that some have labeled it the "new civil right".[14] If young people are to become powerful citizens with full control over their lives then they need to be able to reason mathematically – to think logically, compare numbers, analyze evidence, and reason with numbers.[15] *Business Week* declared that "the world is moving into a new age of numbers" (Jan 23, 2006). Mathematics classrooms need to catch up – not only to help future employers and employees, or even to give students a taste of authentic mathematics, but to prepare young people for their lives.

Engineering is one of the most mathematical of jobs and entrants to the profession need to be proficient in high levels of maths in order to be considered. Julie Gainsburg studied structural engineers at work for over 70 hours and found that although they used mathematics extensively in their work, they rarely used standard methods and procedures. Typically the engineers needed to interpret the problems they were asked to solve (such as the design of a car park or the support of a wall) and form a simplified model to which they could apply mathematical methods. They would then select and adapt methods that could be applied to their models, run calculations, using various representations as they worked, (graphs, words, equations, pictures, and tables) and justify and communicate their methods and results. Thus the engineers engaged in flexible problem solving, adapting and using mathematics, and although they occasionally faced situations when they could simply use standard mathematical formulae this was rare and the problems they worked on were 'usually ill structured and open ended'. As Gainsburg writes: 'Recognizing and defining the problem and wrangling it into a solvable shape are often part of the work; methods for solving have to be chosen or adapted from multiple possibilities, or even invented; multiple solutions are usually possible; and identifying the "best" route is rarely a clear-cut determination, thanks to the competing priorities of the various parties.[16] Gainsburg's findings echo those of other studies of high-level mathematics in use in such areas as design, technology and medicine[17,18] and her conclusion is damning: The traditional mathematics curriculum, 'with its focus on performing computational manipulations, is unlikely to prepare students for the problem-solving demands of the high-tech workplace.'[19]

London University professors Celia Hoyles, and Richard Noss, and their colleagues, performed analyses of mathematics in a range of work settings such as engineering, nursing and banking. In their study of British nurses, Hoyles, Noss and Pozzi[20] (2000) focused upon ratio and proportion, two areas of mathematics that nurses use regularly as they calculate and administer drug doses. Mathematical calculations in nursing need to be conducted with extreme precision, as mistakes could have very serious consequences. Because of the pervasiveness of drug calculations as well as their importance, the researchers found that nurses in different settings had all been taught what was known as the "nursing rule" for working out how much of a drug was needed.

The rule was:

$$\frac{\text{dose prescribed}}{\text{dose per measure}} \times \text{number of measures} = \text{amount of drug needed}$$

For example, if 300 mg of drugs are prescribed and they come in packages of 120mg per 2ml then the volume can be calculated as 300 /120 × 2ml. Hoyles and colleagues found that all of the nurses knew this rule well and could recite it at any time but that they had developed a different way of calculating that served as well or better than the formal rule they had learned. For example, when the nurses needed to perform calculations, (which were always performed error-free) their calculations were shaped by the situation they were in. The researchers termed one approach that they observed 'chunking', which meant that nurses would consider the amount in the dose prescribed and then "chunk" the amounts available to them – so, for example if 60ml was prescribed and they had pills of 15ml concentration, they would chunk the pills into groups of 4. This meant that nurses would combine portions of mass available in standard packs to give the appropriate dose and then conduct parallel calculations on the volumes of solution. The researchers found that the nurses' strategies were structured by familiar aspects of their work, such as packing conventions and typical dosages of the drugs as well as their clinical knowledge of what seemed right for any given drug and patient. This flexible use of mathematical methods, taking into account the particular needs and constraints of the workplace, has proved to be a finding of all of the work studies of everyday mathematics in use.

Studies of "everyday" people using mathematics in their lives – when shopping and performing other more routine tasks – have emerged with similar findings. Researchers have found that adults cope well with mathematical demands, but they draw from school knowledge infrequently. In real world situations, such as street markets and shops, individuals have rarely made use of any school learned mathematical methods or procedures, instead they have created methods that work given the constraints of the situations they faced.[21,22,23,24] Jean Lave, professor from the University of California, Berkeley, found that shoppers used their own methods to work out which were better deals in shops, without using any formal methods learned in school, and that dieters used informal methods that they created when needing to work out measures of servings. For example a dieter who was told he could eat ¾ of a ⅔ cup of cottage cheese did not perform the standard algorithm for multiplying the fractions, instead he emptied ⅔ of a cup onto a measuring board, patted it into a circle marked a cross on it and took one quadrant away leaving ¾ of it.[25] Lave gives many examples from her various studies of people using informal methods such as these, without any recourse to school learned methods.

The ways in which people use mathematics in the world will probably sound familiar to most readers as it is the way many of us use mathematics. Adults rarely stop to remember formal algorithms; instead successful mathematical users size up situations and adapt and apply mathematical methods, using them flexibly.

The maths of the world is so different from the maths that is taught in most classrooms that young people often leave school ill-prepared for the demands of their work and lives. Children learn, even when they are still in school, about the irrelevance of the work they are given, an issue that becomes increasingly important to children as they reach and move through adolescence. As part of a research study in England[26] I interviewed students who had learned traditionally, as well as those who had learned through a problem solving approach, about their use of maths in the part-time jobs that they worked in after school. The students from the traditional approach all said that they used and needed maths out of school but that they would never make use of the mathematics they were learning in their school classrooms. The students regarded the school mathematics classroom as a separate world with clear boundaries that separated it from their lives. The students who had learned through a problem solving approach did not regard the mathematics of school and the world as different and talked with ease about their use of the school mathematics they had learned in their jobs and lives.

We need to bring real maths into maths classrooms and children's lives, instead of the fake version that goes on in many maths classrooms, and we must treat this as a matter of urgency. The suggestions I will give, for things to do, can take place in classrooms or homes, as they are all about being mathematical. Children need to solve ill-structured problems, to ask many forms of questions, to draw and visualize maths and to use, adapt and apply methods. It is also critical that we help children develop self-confidence in maths, even when schools seem to be doing the opposite. Maths is often used by teachers as a tool to sort, track and label children, with many children being told that they are in a mathematical under-class. The brutal labeling of children that goes on in English schools, particularly in maths, puts us out of synch with the rest of the world. This book is aimed to help children engage in the sort of maths that will really help them, as well as giving them opportunities to believe in themselves again.

I conduct longitudinal studies of children's mathematics learning, which are very rare. Typically researchers visit classrooms at a particular point in time to observe children learning, but I have followed students through years of maths classes, in England and America, to observe how their learning develops over time. I am currently the Marie Curie Professor of Mathematics Education at The University of Sussex in the UK, prior to that I was professor at Stanford University, California. In both the UK and the US I have spent years studying thousands of children, following them through middle and secondary schools. In my studies I monitor how children are learning, finding out what's helpful to them and what's not. A few summers ago in California, I returned to middle school classrooms to teach children myself, with some of my doctoral students. The class was a group of largely disaffected students who hated maths and were getting low marks in their maths classes. They started our course saying they didn't want to be there but they ended up loving it, telling us that it had transformed their view of maths. One boy told us that if maths was like that during the school year he would take it all day and every day. One of the girls told us that maths class had always appeared so black and white to her, but in this class it was "all the colours in the rainbow". Our teaching methods were not revolutionary: we talked with the children about maths, and we worked on algebra and arithmetic through puzzles and problems such as the chessboard problem (the answer is not 64):

How many squares are there in a chessboard?

and the 2 jars problem:

Given a five-litre jar and a three-litre jar and an unlimited supply of water, how do you measure out four litres exactly?

Through these and other problems we taught students to use and enjoy mathematics and to reason mathematically.

Successful teachers use teaching methods that more people should know about. Good students also use strategies that make them successful – they are not just people who are born with some sort of maths gene, as many people think. High achieving students are people who learn, whether through great teachers, role models, family, or other sources, to use the particular strategies that I will share in this book.

Based on my studies of thousands of children, *The Elephant in the Classroom* will identify the problems that students encounter and share some solutions. I know that many parents are afraid of maths and don't think that they know enough to help their children, or to talk with them about maths, especially when they start secondary school. But the sorts of mathematical activities that children and parents can do at home don't need a lot of maths knowledge, instead they need a certain approach to maths and learning that I will share in this book.

Maths classrooms should also change for the better, and this book will help equip parents with the knowledge they need to help schools change. Many more children could become successful in mathematics if they learned to approach mathematics *differently*. I hope that this book will spark an interest in people who have been wounded by maths experiences and have hated maths ever since, inspire those who already enjoy maths, and guide those who want to teach children to enjoy maths.

Research in mathematics has produced technologies that have transformed our world with applications in medicine, television, radio, computing, and films, to name but a few, but students typically have no idea of the ways mathematics is used in life, or the relationships that mathematics illuminates. They are not taught to love or even value mathematics and as a result we have insufficient numbers of people to continue important path-breaking scientific and technological work. Something needs to change.

1 / What is Maths?

And why do we all need it?

In my different research studies I have asked hundreds of children, taught traditionally, to tell me what maths is. They will typically say such things as "numbers" or "lots of rules". Ask mathematicians what maths is and they will more typically tell you that it is "the study of patterns" or that it is a "set of connected ideas". Students of other subjects, such as English and science, give similar descriptions of their subjects to experts in the same fields. Why is maths so different? And why is it that students of maths develop such a distorted view of the subject?

Reuben Hersh, a philosopher and mathematician, has written a book called 'What is Mathematics, Really?' in which he explores the true nature of mathematics and makes an important point – people don't like mathematics because of the way it is *mis-represented* in school. The maths that millions of school children experience is an impoverished version of the subject

that bears little resemblance to the mathematics of life or work, or even the mathematics in which mathematicians engage.

What is mathematics, really?

Mathematics is a human activity, a social phenomenon, a set of methods used to help illuminate the world, and it is part of our culture. In Dan Brown's best-selling novel *The DaVinci Code*[1], the author introduces readers to the 'divine proportion,' a ratio that is also known as the Greek letter phi. This ratio was first discovered in 1202 when Leonardo Pisano, better known as Fibonacci, asked a question about the mating behavior of rabbits. He posed this problem:

> A certain man put a pair of rabbits in a place surrounded on all sides by a wall. How many pairs of rabbits can be produced from that pair in a year if it is supposed that every month each pair begets a new pair which from the second month on becomes productive?

The resulting sequence of pairs of rabbits, now known as the Fibonacci sequence, is

$$1, 1, 2, 3, 5, 8, 13, \ldots$$

Moving along the sequence of numbers, dividing each number by the one before it, produces a ratio that gets closer and closer to 1.618, also known as phi, or *the golden ratio*. What is amazing about this ratio is that it exists throughout nature. When flower seeds grow in spirals they grow in the ratio 1.618:1. The ratio of spirals in seashells, pinecones and pineapples is exactly the same. For example, if you look very carefully at the photograph of a daisy below you will see that the seeds in the centre of the flower form spirals, some of which curve to the left and some to the right.

If you map the spirals carefully you will see that close to the centre there are 21 running anti-clockwise. Just a little further out there are 34 spirals running clockwise. These numbers appear next to each other in the Fibonnacci sequence.

Daisy showing twenty-one anti-clockwise spirals

Daisy showing thirty-four clockwise spirals

Remarkably, the measurements of various parts of the human body have the exact same relationship. Examples include a person's height divided by the distance from tummy button to the floor; or the distance from shoulders to finger-tips, divided by the distance from elbows to finger-tips. The ratio turns out to be so pleasing to the eye that it is also ubiquitous in art and architecture, featuring in the United Nations Building, the Greek Parthenon, and the pyramids of Egypt.

Ask most mathematics students in secondary schools about these relationships and they will not even know they exist. This is not their fault of course, they have never been taught about them. Mathematics is all about illuminating relationships such as those found in shapes and in nature. It is also a powerful way of expressing relationships and ideas in numerical, graphical, symbolic, verbal and pictorial forms. This is the wonder of mathematics that is denied to most children.

Those children who do learn about the true nature of mathematics are very fortunate and it often shapes their lives. Margaret Wertheim, a science reporter for *The New York Times*, reflects upon an Australian mathematics classroom from her childhood and the way that it changed her view of the world:

When I was ten years old I had what I can only describe as a mystical experience. It came during a math class. We were learning about circles, and to his eternal credit our teacher, Mr Marshall, let us discover for ourselves the secret image of this unique shape: the number known as pi. Almost everything you want to say about circles can be said in terms of pi, and it seemed to me in my childhood innocence that a great treasure of the universe had just been revealed. Everywhere I looked I saw circles, and at the heart of every one of them was this mysterious number. It was in the shape of the sun and the moon and the earth; in mushrooms, sunflowers, oranges, and pearls; in wheels, clock faces, crockery, and telephone dials. All of these things were united by pi, yet it transcended them all. I was enchanted. It was as if someone had lifted a veil and shown me a glimpse of a marvelous realm beyond the one I experienced with my senses. From that day on I knew I wanted to know more about the mathematical secrets hidden in the world around me.[2]

How many students who have sat through maths classes would describe mathematics in this way? Why are they not enchanted, as Wertheim was, by the wonder of mathematics, the insights it provides into the world, the way it elucidates the patterns and relationships all around us? It is because they are misled by the image of maths presented in school mathematics classrooms and they are not given an opportunity to experience real mathematics. Ask most school students what maths is and they will tell you it is a list of rules and procedures that need to be remembered.[3] Their descriptions are frequently focused on calculations. Yet as Keith Devlin, mathematician and writer of several books about maths points out, mathematicians are often not even very good at calculations as they do not feature centrally in their work. Ask mathematicians what maths is and they are more likely to describe it as *the study of patterns*.[4,5]

Early in his book 'The Math Gene' Devlin tells us that he hated maths in his English primary school. He then recalls his reading of W.W. Sawyer's book 'Prelude to Mathematics' during secondary school that captivated his thinking and even made him start considering becoming a mathematician himself. Devlin quotes the following from Sawyer's book:

'Mathematics is the classification and study of all possible patterns. Pattern is here used in a way that everybody may agree with. It is to be understood in a very wide sense, to cover almost *any kind of regularity that can be recognized by the mind*. Life, and certainly intellectual life, is only possible because there are certain regularities in the world. A bird recognizes the black and yellow bands of a wasp; man recognizes that the growth of a plant follows the sowing of a seed. In each case, a mind is aware of pattern.'[6]

Reading Sawyer's book was a fortunate event for Devlin, but insights into the true nature of mathematics should not be gained *in spite* of school experiences, nor should they be left to the few who stumble upon the writings of mathematicians. I will argue, as others have done before me, that school class-rooms should give children a sense of the nature of mathematics, and that such an endeavor is critical in halting the low achieve-ment and participation that is so commonplace. School children know what English literature and science are because they engage in authentic versions of the subjects in school. Why should mathematics be so different?[7]

What do mathematicians do, really?

Fermat's Last Theorem, as it came to be known, was a theory pro-posed by the great French mathematician, Pierre de Fermat, in the 1630's. Proving (or disproving) the theory that Fermat set out became the challenge for centuries of mathematicians and caused the theory to become known as "the world's greatest mathematical problem."[8] Fermat was born in 1603 and was famous in his time for posing intriguing puzzles and discover-ing interesting relationships between numbers. Fermat claimed that the equation $a^n + b^n = c^n$ has no solutions for n when n is greater than 2 and a non zero integer. So, for example, no num-bers could make the statement $a^3 + b^3 = c^3$ true. Fermat developed his theory through consideration of Pythagoras' famous case of $a^2 + b^2 = c^2$. School children are typically intro-duced to the Pythagorean formula when learning about triangles, as any right-angled triangle has the property that the sum of squares built on the two sides ($a^2 + b^2$) is equal to the square of the hypotenuse c^2. So, for example, when the sides of a triangle are 3 and 4 then the hypotenuse must be 5 because $3^2 + 4^2 = 5^2$. Sets of three numbers that satisfy Pythagoras' case are those where two square numbers (eg 4, 9, 16, 25) can combine to produce a third.

Fermat was intrigued by the Pythagorean triples and explored the case of cube numbers, reasonably expecting that some pairs of cubed numbers could be combined to produce a third cube. But Fermat found this was not the case and the resulting cube always has too few or too many blocks, for example:

A Close Call

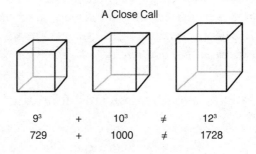

| 9^3 | + | 10^3 | $\neq$ | 12^3 |
| 729 | + | 1000 | $\neq$ | 1728 |

The sum of the volumes of cubes of dimension 9 and 10 almost equals the volume of a cube of dimension 12, but not quite (it is one short!).

Indeed Fermat went on to claim that even if every number in the world was tried, no-one would ever find a solution to $a^3 + b^3 = c^3$ nor to $a^4 + b^4 = c^4$, or any higher power. This was a bold claim involving the universe of numbers. In mathematics it is not enough to make such claims, even if the claims are backed up by hundreds of cases, as mathematics is all about the construction of time-resistant proofs. Mathematical proofs involve making a series of logical statements from which only one conclusion can follow and, once constructed, they are always true. Fermat made an important claim in 1630 but he did not provide a proof and it was the proof of his claim that would elude and frustrate mathematicians for over 350 years. Not only did Fermat not provide a proof but he scribbled a note in the margin of his work saying that he had a "marvelous" proof of his claim but that there was not enough room to write it. This note tormented mathematicians for centuries as they tried to solve what some have claimed to be *the world's greatest mathematical problem*.[9]

'Fermat's last theorem' stayed unsolved for over 350 years, despite the attentions of some of the greatest minds in history. In recent years it was dramatically solved by a shy English mathematician, and the story of his work, told by a number of biographers, captures the drama, the intrigue and the allure of mathematics that is unknown by many. Any child – or adult – wanting to be inspired by the values of determination and per-sistence, enthralled by the intrigue of puzzles and questions, and introduced to the sheer beauty of living mathematics should read Simon Singh's book *Fermat's Enigma*. Singh describes 'one of the greatest stories in human thinking'[10] providing impor-tant insights into the ways mathematicians work.

Many people had decided that there was no proof to be found of Fermat's theorem and that this great mathematical problem was unsolvable. Prizes were offered from different corners of the globe and men and women devoted their lives to finding a proof, to no avail. Andrew Wiles, the mathematician who would write his name into history books, first encountered Fermat's theory as a 10 year old boy while reading in his local library in his home town of Cambridge. Wiles described how he felt when he read the problem, saying that 'It looked so simple, and yet all the great mathematicians in history could not solve it. Here was a problem that I, as a ten-year-old, could understand and I knew from that moment that I would never let it go, I had to solve it.'[11] Years later Wiles graduated with a PhD in mathematics from Cambridge and then moved to Princeton to take a position in the mathematics department. But it was still some years later when Wiles realized that he could devote his life to the problem that had intrigued him since childhood.

As Wiles set about trying to prove Fermat's Last Theorem he retired to his study and started reading journals, gathering new techniques. He started exploring and looking for patterns, working on small areas of mathematics and then standing back to see if they could be illuminated by broader concepts. Wiles worked on a number of different techniques over the next few years, exploring different methods for attacking the problem. Some seven years after starting the problem Wiles emerged from his study one afternoon and announced to his wife that he had solved Fermat's Last Theorem.

The venue that Wiles chose to present his proof of the 350 year-old problem was a conference at the Isaac Newton Institute in Cambridge, England in 1993. Some people had become intrigued about Wiles' work and rumours had started to filter through that he was actually going to present a proof of Fermat's Last Theorem. By the time Wiles came to present his work there were over two hundred mathematicians crammed into the room, and some had sneaked in cameras to record the historic event. Others – who could not get in – peered through windows. Wiles needed three lectures to present his work and at the conclusion of the last lecture the room erupted into great applause. Singh described the atmosphere of the rest of the conference as 'euphoric' with the world's media flocking to the Institute. Was it possible that this great and historical problem had finally been solved? Barry Mazur, a number theorist and algebraic geometer, reflected on the event saying that 'I've never seen such a glorious lecture, full of such wonderful ideas, with such dramatic tension, and what a build up. There was only one possible punch line.' Everyone who had witnessed the event thought that Fermat's Last Theorem was finally proved. Unfortunately, there was an error in Wiles' proof that meant that Wiles had to plunge himself back into the problem. In

September 1994, after a few more months of work, Wiles knew that his proof was complete and correct. Using many different theories, making connections that had not previously been made, Wiles had constructed beautiful new mathematical methods and relationships. Ken Ribet, a Berkeley mathematician whose work had contributed to the proof, concluded that the landscape of mathematics had changed and mathematicians in related fields could work in ways that had never been possible before.

The story of Wiles is fascinating and told in more detail by Simon Singh and others. But what do such accounts tell us that could be useful in improving children's education? One clear difference between the work of mathematicians and school-children is that mathematicians work on long and complicated problems that involve combining many different areas of mathematics. This stands in stark contrast to the short questions that fill the hours of maths classes and that involve the repetition of isolated procedures. Long and complicated problems are important to work on for many reasons, one of them being that they encourage persistence, one of the values that is critical for young people to develop and that will stand them in good stead in life and work. When mathematicians are interviewed they often speak of the enjoyment they experience from working on difficult problems. Diane Maclagan, a professor at Rutgers University in the US, was asked: what is the most difficult aspect of your life as a mathematician? She replied "Trying to prove theorems". And the most fun? the interviewer asked. "Trying to prove theorems." She replied.[12] Working on long and complicated problems may not sound like fun, but mathematicians find such work enjoyable because they are often successful. It is hard for any school child to enjoy a subject if they experience repeated failure, which of course is the reality for many young people in school mathematics classrooms. But the reason that mathematicians are successful is because they have learned something very important – and very learnable. They have learned to problem solve.

Problem solving is at the core of mathematician's work, as well as the work of engineers and others, and it starts with the making of a guess. Imre Lakatos, mathematician and philosopher, describes mathematical work as 'a process of "conscious guessing" about relationships among quantities and shapes'[13]. Those who have sat in traditional maths classrooms are probably surprised to read that mathematicians highlight the role of guessing, as I doubt whether they have *ever* experienced any encouragement to guess in their maths classes. When an official report in the UK was commissioned to examine the mathematics needed in the workplace the reviewers found that estimation was the most useful mathematical activity.[14] Yet when children who have experienced traditional maths classes are asked to estimate they are often completely flummoxed and try to work out exact answers then round them off to look like an estimate. This is because they have not developed a good *feel* for numbers, which would allow them to estimate instead of calculate, and also because they have learned, wrongly, that mathematics is all about precision, not about making estimates or guesses. Yet both are at the heart of mathematical problem solving.

After making a guess mathematicians engage in a zig-zagging process of conjecturing, refining with counter-examples, and then proving. Such work is exploratory and creative and many writers draw parallels between mathematical work and art or music. Robin Wilson, a British mathematician, proposes that mathematics and music 'are both creative acts. When you are sitting with a bit of paper creating mathematics, it is very like sitting with a sheet of music paper creating music.'[15] Devlin agrees saying that 'Mathematics is not about numbers, but about life. It is about the world in which we live. It is about ideas. And far from being dull and sterile, as it is so often portrayed, it is full of creativity.'[16]

The exhilarating, creative pathways that mathematicians describe as they solve problems, often hidden in the end-point of mathematical work, cannot be the exact same pathways that school children experience, as children need to be taught the methods they need, as well as use them in the solving of problems, but neither should school mathematics be so different as to be unrecognizable. As George Pólya, the eminent Hungarian mathematician, reflected, in 1945:

> 'A teacher of mathematics has a great opportunity. If he fills his allotted time with drilling his students in routine operations he kills their interest, hampers their intellectual development, and misuses his opportunity. But if he challenges the curiosity of his students by setting them problems proportionate to their knowledge, and helps them to solve their problems with stimulating questions, he may give them a taste for, and some means of, independent thinking.'[17] (Polya, 1971, v)

Another interesting feature of the work of mathematicians is its collaborative nature.

Many people think of mathematicians as people who work in isolation, but this is far from the truth. Leone Burton, a professor of mathematics education, interviewed 70 research mathematicians and found that they generally challenged the solitary stereotype of mathematical work, reporting that they preferred to collaborate in the production of ideas. Over half of the papers they submitted to Burton as representative of their work were written with colleagues. The mathematicians interviewed gave many reasons for collaboration, including the advantage of learning from one another's work, increasing the quality of ideas, and sharing the 'euphoria' of problem solving. As Burton reflected, 'they offered all the same reasons for collaborating on research that are to be found in the educational literature advocating collaborative work in classrooms.'[18] Yet non-collaborative maths classrooms continue to prevail across England.

Something else that we learn from various accounts of mathematicians' work is that an important part of real, living mathematics is the posing of problems. Viewers of *A Beautiful Mind* may remember John Nash (played by Russell Crowe) undergoing an emotional search to form a question that would be sufficiently interesting to be the focus of his work. People commonly think of mathematicians as solving problems but as Peter Hilton[19], an algebraic topologist, has said 'Computation involves going from a question to an answer. Mathematics involves going from an answer to a question.' Such work requires creativity, original thinking, and ingenuity. All the mathematical methods and relationships that are now known and taught to school children started as questions, yet students do not see the questions. Instead they are taught content that often appears as a long list of answers to questions that nobody has ever asked. Reuben Hersh, an American mathematician, puts it well:

'The mystery of how mathematics grows is in part caused by looking at mathematics as answers without questions. That mistake is made only by people who have had no contact with mathematical life. It's the questions that drive mathematics. Solving problems and making up new ones is the essence of mathematical life. If mathematics is conceived apart from mathematical life, of course it seems – dead.'[20]

Bringing mathematics back to life for school children involves giving them a sense of living mathematics. When school students are given opportunities to ask their own questions and to extend problems into new directions, they know mathematics is still alive, not something that has already been decided and just needs to be memorized. Posing and extending problems of interest to students mean they enjoy mathematics more, they feel more ownership of their work and they ultimately learn more. English school children used to work on long problems that they could extend into directions that were of interest to them in maths classes. For example, in one problem students were asked to design any type of building. This gave them the opportunity to consider interesting questions involving high-level mathematics, such as the best design for a fire station with a firefighter's pole. Teachers used to submit the students' work to examination boards and it was assessed as part of the students' final grade. When I asked English school children about their work on these problems they not only reported that they were enjoyable and they learned a lot from them, but that their work made them "feel proud" and that they could not feel proud of their more typical textbook work. Mathematical coursework no longer exists in England as it was decided that it too often led to cheating. Unfortunately this was one of few experiences schoolchildren had to use maths in the solving of real and interesting problems.

Another important part of the work of mathematicians that enables successful problem solving is the use of a range of representations such as symbols, words, pictures, tables and diagrams, all used with extreme precision. The precision required in mathematics has become something of a hallmark for the subject and it is an aspect of mathematics that both attracts and repels. For some school children it is comforting to be working in an area where there are clear rules for ways of writing and communicating. But for others it is just too hard to separate the precision of mathematical language with the uninspiring "drill and kill" methods that they experience in their maths classrooms. There is no reason that precision and drilled teaching methods need to go together and the need for precision with terms and notation does not mean that mathematical work precludes open and creative exploration. On the contrary, it is the fact that mathematicians can rely on the precise use of language, symbols and diagrams that allows them to freely explore the *ideas* that such communicative tools produce. Mathematicians do not play with the notations, diagrams, and words as a poet or artist might, instead they explore the relations and insights that are revealed by different arrangements of the notations. As Keith Devlin reflects:

'Mathematical notation no more *is* mathematics than musical notation is music. A page of sheet music represents a piece of music, but the notation and the music are not the same; the music itself happens when the notes on the page are sung or performed on a musical instrument. It is in its performance that the music comes alive; it exists not on the page but in our minds. The same is true for mathematics.'[21]

Mathematics is a performance, a living act, a way of interpreting the world. Imagine music lessons in which students worked through hundreds of hours of sheet music, adjusting the notes on the page, receiving ticks and crosses from the teachers, but never playing the music. Students would not continue with the subject because they would never experience what music *was*. Yet this is the situation that continues in mathematics classes, seemingly unabated.

Those who use mathematics engage in mathematical *performances*, they use language in all its forms, in the subtle and precise ways that have been described, in order to do something with mathematics. Students should not just be memorizing past methods; they need to engage, do, act, perform, problem solve, for if they don't *use* mathematics as they learn it they will find it very difficult to do so in other situations, including examinations.

The erroneous thinking behind many school approaches is that students should spend years being drilled in a set of methods that they can use later. Many mathematicians are most concerned about the students who will enter post-graduate programs in mathematics. At that point students will encounter real mathematics and use the tools they have learned in school to work in new, interesting and authentic ways. But by this time most maths students have given up on the subject. We cannot keep pursuing an educational model that leaves the best and the only real taste of the subject to the end, for the rare few who make it through the grueling eleven years that precede it. If students were able to work in the ways mathematicians do, for at least some of the time – posing problems, making guesses and conjectures, exploring with and refining ideas, and discussing ideas with others, then they would not only be given a sense of true mathematical work, which is an important goal in its own right,[22] they would also be given the opportunities to enjoy mathematics and learn it in the most productive way.[23,24,25]

2 / What's going wrong in classrooms?

Identifying the problems.

Targets, Levels and Labels.

One of the biggest problems with maths teaching in England is the desire of many maths teachers to label children, assign them a level, and prejudge their achievement. Too many maths teachers think that their role is to find the chosen few who are really good at maths, assigning the rest to low level sets and giving them low level work for the rest of their school lives. Schools in England have always had this tendency but the last ten years of government directives, with instructions to target and label children in order to track their progress, have pushed the situation out of control. Schools are now deciding which children can *and cannot* do maths when they are only 4 years old, and research has shown that when children get put into low sets 88% of them stay there until they leave school[1]. The fact that our children's future is decided for them at an early age derides the work of schools and contravenes basic knowledge about child development and learning. Children develop at different rates, and they reveal different interests, strengths and dispositions at various stages of their development. One of the most important goals of schools is to provide stimulating environments for all children; environments in which children's interest can be peaked and nurtured, with teachers who are ready to recognize, cultivate and develop the potential that children show at different times and in different areas. It is difficult to support a child's development and nurture their potential if they are placed into a low group at an early age, told that they are achieving at lower levels than others, given less challenging and interesting work, taught by less qualified and less experienced teachers, and separated from peers who would stimulate their thinking. The way that we treat maths learners in England is hugely problematic and it separates the educational system from the rest of the World.

Over the last ten years, teachers have been subject to a torrent of government directives telling schools to target, label and test children endlessly. One of the results of our target culture is that maths teaching has been reduced to a *can or can't do* exercise. Teachers and students alike seem to believe that the purpose of maths is to sort those who can, from those who can't. It seems that students no longer learn maths in order to develop problem-solving skills, or to appreciate patterns and connections, or even to express relationships and ideas using numerical, pictorial, verbal and symbolic forms. Now students learn maths so that they can jump through the right hoops and achieve their "target grades". It is no coincidence that the last eight years, in which target setting has taken over schools, are the same eight years that the UK has dropped from eighth to twenty-fourth place in international tests of mathematical problem solving[2]. The target driven, can and can't do culture that pervades schools is not good for anyone. It might seem good to be one of the few "can do" students in maths, but recent research has shown that high achievers in maths often go through school thinking they should choose maths for further study just because they are one of the few people that can do it. Unfortunately when they arrive at University they find that they are surrounded by other students who are just as good as they are and they start to question their whole purpose for being there[3]. If they are not one of the few who "can do" what is the point? This is particularly true if they have learned mathematics without ever developing an appreciation for its beauty or applicability. And things are even worse for the "can't do" children who fill maths classrooms nationwide. It is critical that we replace this target driven, can or can't do culture with a return to learning for its own sake. Children need to be taught within a culture that values their ability to think and reason, invites their engagement in the breadth of mathematics, and encourages all students to achieve at high levels, even those who struggle for some of the time.

The drive to label and prejudge children in maths puts England out of synch with the rest of the world. In countries where students are mathematical high achievers, such as Finland and Japan, schools communicate the idea that everyone can be good at maths and teachers work to make sure that happens. In England teachers often communicate the opposite, from an early age. Our common practice of telling students they are "low ability" is a major reason that children from the UK have the lowest sense of well being of children in 20 countries[4]. Chapters 3 and 5 will explain, in detail, the best ways to group children and to counter the negative messages that may be given by schools. Chapter 9 will talk about ways to work with schools to help them understand the most effective, label-free ways of teaching students mathematics.

Learning without thought?

In some educational debates, traditional teaching is placed at one end of a pole, and progressive teaching at the other. In my research I have found that such categories do not mean much, and that both camps include effective and ineffective teaching. Certain teachers might be described as traditional because they lecture and students work individually, but they also ask students great questions, engage them in interesting mathematical inquiries and give students opportunities to solve problems, not just rehearse standard methods. Such teachers are wonderful and I wish there were many more of them. The type of 'traditional' teaching that concerns me greatly and that I have identified from decades of research as highly ineffective is a version that encourages *passive learning*.

In many mathematics classrooms across England the same ritual unfolds: teachers stand at the front of class demonstrating methods for 20-30 minutes of class time each day while students copy the methods down in their books, then students work through sets of near-identical questions, practicing the methods. Students in such classrooms quickly learn that *thought* is not required in maths class and that the way to be successful is to watch the teacher carefully and copy what they do. In interviews with hundreds of students from such classes I have asked them what it takes to be successful in maths class and they almost always give the exact same answer: you have to *pay careful attention*. As one of the girls I interviewed told me: *In maths you have to remember; in other subjects you can think about it.*

Students taught through passive approaches follow and memorize methods instead of learning to inquire, ask questions and solve problems. I have interviewed hundreds of students taught in such ways. This is Sue's reaction to her maths teaching, a reaction shared by many children:

I'm just not interested in, just, you give me a formula, I'm supposed to memorize the answer, apply it and that's it.

Sanjit went on to explain why such models of teaching held no interest for him:

You have to be willing to accept that sometimes things don't look like – they don't seem that you should do them. Like they have a point. But you have to accept them.

Students are forced into a passive relationship with their knowledge – they are taught only to follow rules and not to engage in sense-making, reasoning, or thought, acts that are critical to an effective use of mathematics. This passive approach, that characterizes maths teaching in many schools, is highly ineffective.

When students try to memorize hundreds of methods, as students do in classes that use a passive approach, they find it extremely hard to use the methods in any new situations, often resulting in failure on exams as well as in life. The secret that good mathematics users know is that only a few methods need to be memorized, and that most mathematics problems can be tackled through the understanding of mathematical concepts and active problem solving.

I have spent my research career conducting unusual studies of learning. They are unusual because instead of dropping in on students to see what they are doing in maths classes, I have followed students through years of secondary school, performing *longitudinal* studies. I have spent thousands of hours, with teams of doctoral students, collecting data on students learning mathematics in different ways. This has included watching hundreds of hours of maths classes, interviewing students about their experiences, giving them questionnaires on mathematical beliefs, and performing assessments to probe students' understanding. These studies have revealed that many maths classrooms leave students cold, disinterested or traumatized. In hundreds of interviews with students who have experienced passive approaches students have told me that thought is not required, or even *allowed*, in maths class. Children emerge from passive approaches believing that they only have to be obedient and memorize what the teacher tells them to do. They learn that they must simply memorize methods even when methods do not make sense. It is ironic that maths, a subject that should be all about inquiring, thinking and reasoning is one that students have come to believe requires *no thought*.

The fact that students are drilled in methods and rules that do not make sense to them is not just a problem for their understanding of mathematics. Such an approach leaves students frustrated, because most of them want to understand what they are learning. Students want to know how different mathematical methods fit together and why they work. This is especially true for girls and women, as I shall explain in chapter 6. The following response from Kate, a girl taking calculus in a traditional class, is one that is similar to those I have received from many young people I have interviewed:

> We knew how to do it. But we didn't know why we were doing it and we didn't know how we got around to doing it. Especially with limits, we knew what the answer was, but we didn't know why or how we went around doing it. We just plugged into it. And I think that's what I really struggled with – I can get the answer, I just don't understand why.

Young people are naturally curious and their inclination – at least before they experience traditional teaching – is to make sense of things and to understand them. Many maths classes rid students of this worthy inclination. Kate was at least fortunate to still be asking *why?* even though she, like others, was not given opportunities to understand why the methods worked. Children begin school as natural problem solvers and many studies have shown that students are better at solving problems *before* they attend maths classes.[5,6] They think and reason their way through problems, using methods in creative ways, but after a few hundred hours of passive maths learning students have their problem solving abilities knocked out of them. They think that they need to remember the hundreds of rules they have practiced and they abandon their common sense in order to follow the rules.

Consider for a moment this mathematics problem:

A woman is on a diet and goes into a shop to buy some turkey slices. She is given 3 slices which weigh ⅓ of a pound but her diet says that she is only allowed to eat ¼ of a pound. How much of the 3 slices she bought can she eat while staying true to her diet?

This is an interesting problem and I urge readers to try it before moving on. This was a problem that was posed by Ruth Parker, a wonderful teacher of teachers who has spent many years working with parents to help them understand the benefits of inquiry approaches. In one of her public sessions with children and parents she posed this problem and asked people to solve it. Her purpose in doing so was to see what kind of solutions people offered and how these compared to their school experiences. Many of the adults who had experienced passive approaches were unable to solve the problem because they could not apply a rule they had learned. Some of them tried ¼ × ⅓, as they knew that something should be multiplied, but they recognized that their answer of ¹⁄₁₂ was probably incorrect. Some tried ¼ × 3 but their answer of ¾ of a pound also did not make sense. To use a rule they needed to set up the following equation:

$$3 \text{ slices} = \tfrac{1}{3}$$
$$x \text{ slices} = \tfrac{1}{4}$$

Once Ruth told them this then the people who had remembered rules and methods were able to do the rest – to cross-multiply and say that:

$$\tfrac{1}{3}x = \tfrac{1}{4}$$
$$\text{so } x = \tfrac{3}{4} \text{ slices}$$

But as she pointed out the most important part of the mathematics that is needed is to be able to set up the equation. This is something that children get very little experience of – they either use the same equation over and over again in a maths lesson and so do not focus on how to set them up, or they are given equations that are already set up for them and they rehearse how to solve them, over and over again.

But look at some of the wonderful solutions offered by young children who had not yet been subjected to rule-bound, passive approaches at school:

One year 5 student said:

If 3 slices is ⅓ of a pound then 9 slices is a pound. I can eat ¼ of a pound so ¼ of 9 slices is ¾ slices.

Another solved the problem visually:
Representing a pound:

And then a quarter of a pound:

These elegant solutions are the sorts of methods that are suppressed by passive, rule-bound mathematics approaches that teach only one way to solve problems and discourage all others. We can only speculate as to whether these same young people would be able to think of solutions like these after future years of passive mathematics approaches. The fact that many students learn to suppress their thoughts, ideas and problem solving abilities in maths classes is one of the most serious indictments of our education system.

Learning without Talking?

Another major problem with passive approaches to mathematics is that students don't talk about maths. Some may believe that working in silence is the optimum learning condition, but in fact this is far from the truth. I have visited hundreds of classrooms in which students sit at their desks silently watching the teacher demonstrate methods and then practice the methods – often chatting to their friends as they do so, but not about the maths. This approach is flawed, for a number of reasons. One problem is that students often need to talk through methods to know whether they really understand them or not. Methods can *seem* to make sense when people hear them, but explaining them to someone else is the best way to know whether they are really understood.

When two famous mathematicians, from very different

circumstances, reflected on the conditions that allowed them to succeed in mathematics, I was struck by the similarity in their statements. Sarah Flannery is a young Irish woman who won the European Young Scientist of the Year award for the development of a 'breathtaking' mathematical algorithm. In her autobiography she writes about the different conditions that promoted her learning, including the 'simple maths puzzles' that she worked on as a child, which I shall talk more about in chapter eight. Flannery writes: 'The first thing I realized about learning mathematics was that there is a hell of a difference between, on the one hand, *listening* to maths being talked about by somebody else and thinking that you are understanding, and, on the other, thinking about maths and understanding it yourself and *talking* about it to someone else.'[7] Reuben Hersh is an American mathematician who wrote the book 'What is Mathematics Really?' In it he also talks about the source of his mathematical understanding, saying that 'mathematics is learned by computing, by solving problems, and by *conversing*, more than by reading and *listening*.'[8]

Both of these successful mathematicians highlight the role of talking over listening, yet listening is the signature of the passive mathematics approaches that are the norm for maths students. The first, faulty learning condition that Flannery describes: 'listening to maths being talked about by someone else' is the quintessence of passive maths approaches. The second condition that enabled her to understand – 'thinking about maths' and 'talking about it to someone else' is what students should be doing in classrooms and homes and is essential to the active approach I will set out in chapter three. When students listen to someone laying out mathematical facts, a passive act that does not necessarily involve intellectual engagement, they usually think that it makes sense but such thoughts are very different from understanding, as the following true story illustrates.

When I was working in the United States I had been visiting a school that had recently moved from problem solving maths lessons to a very traditional approach, because of campaigning by external conservative groups[9]. Soon after the students had been put into classes called "traditional math" I visited a class of thirteen year olds. They were being taught by an "old school" maths teacher who strode up and down at the front of the room filling the board with mathematical methods that he explained to students. He joked occasionally and peppered his sentences with phrases such as "this is easy" and "just do this quickly". The students liked him, because of his jokes and cheery outlook, and his clear explanations, and they would watch and listen carefully and then practice the methods in their books. The students had all been involved in the very public debate at the school between those supporting a "traditional" maths approach and those supporting newer curriculum materials. One day when I was visiting the algebra class that the students knew as "traditional math", I stopped and knelt by the side of one of the boys and asked him how he was getting on. He replied enthusiastically saying "Great, I love traditional math, the teacher tells it to you and you get it". I was pleased that he was so positive and was about to go to another desk when the teacher came around handing tests back. The boy's face fell as he saw a large F circled in red on the front of his test. He looked at the F, looked through his test and turned back to me saying "of course that's what I hate about traditional math, you think you've got it when you haven't!" This after-thought, given with a wry smile, was amusing, but it also communicated something very important about the limitations of the approach that he was experiencing. Students do think that they "get it" when methods are shown to them, clearly, on the board and they repeat them lots of times but there is a huge difference between seeing something that appears to make sense and understanding it well enough to use it a few weeks or days later, or in different situations. To know whether students are understanding methods as opposed to just thinking that everything makes sense, they need to be solving problems – not just repeating procedures with different numbers – and they need to be talking through and explaining different methods.

Another problem with the silent approach is that it gives students the wrong idea about mathematics. One of the most important parts of being mathematical is an action called *reasoning*. This involves explaining *why* something makes sense and how the different parts of a mathematical solution lead on from one another. It is very hard to reason about mathematics when working in silence. Students who learn to reason and to *justify* their solutions are also learning that mathematics is about making sense. Whenever students offer a solution to a maths problem, they should know why the solution is appropriate, and they should draw from mathematical rules and principles when they justify the solution rather than just saying that a textbook or a teacher told them it was right. Reasoning and justifying are both critical mathematical acts and it is very difficult to engage in them without talking. If students are to learn that being mathematical involves making sense of their work and being able to explain it to someone else, justifying the different moves, then they need to talk to each other and to their teacher.

Another reason that talking is so critical in mathematics classrooms is that when students discuss maths they come to know that the subject is more than a collection of rules and methods set out in books – they realize that mathematics is a subject that they can have their own ideas about; a subject that can invoke different perspectives and methods; and a subject that is connected through organizing concepts and themes. This is important for all learners but perhaps none more so than adolescents. If young people are asked to work in silence and they are not asked to offer their own ideas and perspectives they often feel disempowered and disenfranchised, ultimately choosing to leave mathematics even when they have performed well.[10] When students are asked to give their ideas on mathematical problems they feel that they are using their intellect and that they have responsibility for the direction of their work, which is extremely important for adolescents.

Mathematical discussions are also an excellent resource for student understanding. When students explain and justify their work to each other, they get to hear each other's explanations, and there are times when students are much more able to understand an explanation from another student than from a teacher. The students who are talking are able to gain deeper understanding through explaining their work, and the ones listening are given greater access to understanding. One of the reasons for this is that when we verbalize mathematical thoughts we need to reconstruct them in our minds, and when others react to them we reconstruct them again. This act of reconstruction deepens understanding.[11] When we work on mathematics in solitude there is only one opportunity to understand the mathematics. Of course discussions need to be organized well and I will explain in later chapters how disciplined and successful discussions are managed in classrooms.

The bottom line on talking is that it is critical to maths

learning and to giving students the depth of understanding they need. This does not mean that students should be talking all the time, or that just any form of talking is helpful. Maths teachers need to organize productive mathematical discussions and they should give children some time to discuss maths and some time to work alone. But the distorted version of mathematics that is conveyed in silent maths classrooms is one that makes mathematics inaccessible and extremely boring to most children.

Learning without reality.

A problem with both old and newer mathematics approaches that has emerged from my research is the ridiculous problems that are used in mathematics classrooms. Just like stepping through the wardrobe door and entering *Narnia*, in maths classrooms trains travel towards each other on the same tracks and people paint houses at identical speeds all day long. Water fills baths at the same rate each minute, and people run around tracks at the same distance from the edge. To do well in maths class, children know that they have to suspend reality and accept the ridiculous problems they are given. They know that if they think about the problems and use what they understand from life then they will fail. Over time, schoolchildren realize that when you enter *Mathsland* you leave your commonsense at the door.

Contexts started to become more common in maths problems in the nineteen-seventies and eighties; up until that point most mathematics had been taught though abstract questions with no reference to the world. The abstractness of mathematics is synonymous for many people with a cold, detached, remote body of knowledge. Some believed that this image may be broken down through the use of contexts, and so mathematics questions were placed into contexts with the best of intentions. But instead of giving students realistic situations that they could analyze, text-book authors began to fill books with make-believe contexts – contexts that students were meant to believe but for which they should not use any of their real world knowledge. Students are frequently asked to work on questions involving, for example, the price of food and clothes, the distribution of pizza, the num-bers of people who can fit into a lift, and the speeds of trains as they rush towards each other, but they are not meant to use any of their actual knowledge of clothing prices, people or trains. Indeed if they do engage in the questions and use their real world knowledge they will fail. Students come to know this about maths class, they know that they are entering a realm in which common-sense and real world knowledge are not needed.

Here are the sorts of examples that fill maths books:

Alan can do a job in 6 hours and Colin can do the same job in 5 hours. What part of the job can they finish by working together for 2 hours?

A restaurant charges £2.50 for one eighth of a quiche. How much does a whole quiche cost?

A pizza is divided into fifths for 5 friends at a party. Three of the friends eat their slices but then 4 more friends arrive. What fractions should the remaining 2 slices be divided into?

Everybody knows that people work together at a different pace than when they work alone, that food sold in bulk is usu-ally sold at a different rate and that if extra people turn up at a party more pizza is ordered or people go without slices – but none of this matters in *Mathsland*. One long-term effect of working on make believe contexts is that such problems con-tribute to the mystery and other-worldliness of *maths* which curtails people's interest in the subject. The other effect is that students learn to ignore contexts and work only with the numbers, a strategy that would not apply to any real world or professional situation. An illustration of this phenomenon is given by this famous question, asked in a national assessment in the US. Students were asked:

An army bus holds 36 soldiers. If 1128 soldier are being bussed to their training site, how many buses are needed?

The most frequent response from students was 31 remainder 12, a non-sensical answer when dealing with the numbers of buses needed.[12] Of course the test writers wanted the answer of 32, and the '31 remainder 12' response is often used as evidence that American students do not know how to interpret situations. But it may equally be given as evidence that they have been trained in *Mathsland* where such responses are sensible.

My argument against pseudo contexts does not mean that contexts should not be used in mathematics examples, as they can be extremely powerful. But they should only be used when they are realistic, and when the contexts offer something to the students, such as increasing their interest or modeling a mathematical concept. A realistic use of context is one where students are given real situations that need mathematical analysis, for which they do need to consider the variables, rather than ignore them. For example, students could be asked to use mathematics to predict population growth. This would involve students interpreting newspaper data on the British population, investigating the amount of growth over recent years, determining rates of change, building linear models ($y=mx+b$) and using these to predict population growth into the future. Such questions are excellent ways to interest students, motivate them and give them opportunities to use mathematics to solve problems.

Contexts may also be used to give a visual representation, helping to convey meaning. It does not hurt to suggest that a circle is a pizza that needs dividing into fractions, but it does hurt when students are invited into the world of parties and friends while at the same time being required to ignore everything they know about parties and friends.

There are also many wonderful problems in mathematics that can engage students with no context, or barely any context at all. Problems such as the famous Four Colour Problem, that intrigued mathematicians for centuries, is a good example of a gripping, abstract maths problem. The problem came about in 1852 when Francis Guthrie was trying to colour a map of the counties of England. He did not want to colour any adjacent counties with the same colour and noticed that he only needed 4 colours to cover the map. Mathematicians then set out to prove that only 4 colours would be needed in any map or any set of touching shapes. It took centuries to prove this and some still question the proof.

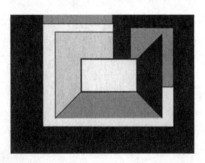

This is a great problem that can be given to students to investigate, they can work with a map of touching countries, such as Europe, or draw their own shapes. For example:

Can you colour this map using only 4 colours, with no two adjacent 'countries' coloured the same?

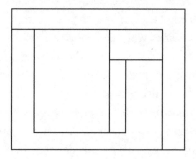

Other examples of problems I have used in this book – the 3 jars problem, or the chessboard problem, use contexts sensibly and responsibly; the contexts give meaning to the problems and provide realistic constraints – students do not have to partly believe them and partly ignore them.

The artificiality of mathematics contexts may seem to be a small concern, but the long-term impact of such approaches can be devastating for students of mathematics. Hilary Rose, a sociologist and newspaper columnist, illustrates this point well. She recalls that as a young child she had been something of 'a mathematical wunderkind'[13] and loved to explore patterns, numbers and shapes. She describes how her sense of mathematical magic ended when 'real world' problems were used in school. At first she enthusiastically engaged with the problems, drawing upon her knowledge of the situations described, but then found that such engagement was not allowed:

'Thinking about it, it was those so called practical problems that irritated me the most. It was obvious to me that many of the questions simply indicated that the questioner did not know enough about the craft skills involved in real world solutions. Lawn rollers being pulled up slopes, wallpapering rooms by calculating square feet and inches: these were tedious and as far as that highly practical child could see, stupid. (. . .) I know that the price I paid was to lose my sense of confidence that school maths and everyday maths were part of one world.'[14]

If the ridiculous contexts were taken out of school maths books the books would probably reduce in size by more than 50%. The elimination of ridiculous contexts would be good for many reasons, most importantly, students would realize that they are learning an important subject that helps make sense of the world, rather than a subject that is all about mystification and non-sense.

As the world changes and technology becomes more and more pervasive in our jobs and lives it is impossible to know exactly which mathematical methods will be most helpful in the future. This is why it is so important that schools develop flexible thinkers who can draw from a variety of mathematical principles in solving problems. The only way to create flexible mathematical thinkers is to give children experience of working in these ways, in school and at home. In the next chapter I will consider four different school approaches, two of which were extremely effective in achieving this goal.

3 / A Vision for a Better Future

Effective Classroom Approaches.

Imagine if there was a time when children were eager to go to maths lessons at school, excited to learn new mathematical ideas, and able to use maths to solve problems outside of the classroom. Adults would feel comfortable with maths, be happy to be given mathematical problems at work, and refrain from saying at parties "I am terrible at maths". We would have the number and range of people good at maths to fill the various jobs needing mathematical and scientific understanding that our technological age requires. All of this might sound far-fetched given the number of mathematically damaged and phobic people in the population and the scores of school children who dread maths lessons. But a very different mathematical reality is possible and achievable, and parents, as well as teachers, have a critical role to play in bringing this reality about. In the pages that follow I will describe two approaches that offered students an experience of real mathematical work and were highly successful. One of the schools I will describe, that used a very unusual and interesting mathematics approach, was in California. The other school that also used a highly successful approach, was in England. Both of the schools are secondary schools but the principles of the two approaches apply to primary and secondary schools. In the following pages we will learn about students from a wide range of backgrounds who came to love maths, to achieve at high levels, and to view mathematics as an important part of their future.

1. The Project Based Approach.

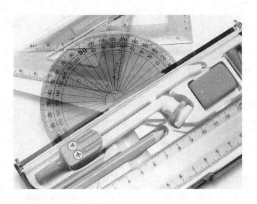

Phoenix Park School

The day that I walked into Phoenix Park school[1], in a working class area of the country, I didn't know what to expect. I had invited the maths department at the school to be a part of my research project. I knew that the department used a "project based approach" but I did not know much more than that. I made my way across the playground and into the school buildings on that first morning with some trepidation. A group of students congregated outside their maths classroom at break time and I asked them what I should expect from the lesson I was about to see. "Chaos" said one of the students, "freedom" said another. Their descriptions made me even more curious about the lessons. Some three years later, after moving with the students through school, observing hundreds of lessons and researching the students' learning, I knew exactly what they meant.

As part of my longitudinal research project on different ways of learning mathematics, I followed an entire cohort of students in each of two schools, from when they were 13 to when they were 16 years of age. I watched hundreds of hours of lessons, interviewed and gave surveys to students and performed various assessments to monitor the ways the different teaching approaches impacted students' learning. One of the schools, Phoenix Park, used a project based approach, the other, Amber Hill, used the more typical, traditional approach. The two schools were chosen because of their different approaches, but also because the student intakes were demographically very similar, the teachers were well qualified and the students had followed exactly the same mathematics approaches up to the age of 13, when my research began. At that time, the students at the two schools scored at the same levels on national mathematics tests. Then their mathematical pathways diverged.

The classrooms at Phoenix Park did look chaotic. The "project based" approach meant a lot less order and control than in traditional approaches. Instead of teaching procedures that students would practice, the teachers gave the students projects to work on that *needed* mathematical methods. From the beginning of year 9 (when students started at the school) to three-quarters of the way through year 11, the students worked on open-ended projects in every lesson. The students were taught in mixed-ability groups and they did not move into groups that were organized around attainment until just before exam time in year 11. Projects usually lasted for about three weeks.

At the start of the different projects the teachers would introduce students to a problem or a theme that the students explored, using their own ideas and the mathematical methods that they were learning. The problems were usually very open so that students could take the work in directions of interest to them and they could all be challenged at an appropriate level. For example, in *volume 216* the students were simply told that the volume of an object was 216 and they were asked to go away and think about what the object could be, what dimensions it would have and what it would look like. Sometimes teachers taught the students mathematical content that could be useful to them before they started a new project. More typically though, the teachers would introduce methods to individuals or small groups when they encountered a need for them within the particular project on which they were working. Simon and Philip described the schools' maths approach in this way:

> S: We're usually set a task first and we're taught the skills needed to do the task, and then we get on with the task and we ask the teacher for help.
> P: Or you're just set the task and then you go about it in . . . you explore the different things, and they help you in doing that.. so you sort of . . . so different skills are sort of tailored to different tasks.
> JB: And do you all do the same thing?
> P: You're all given the same task, but how you go about it, how you do it and what level you do it at, changes, doesn't it? (Simon and Philip, PP, year 10, JC)

The students were given an unusual degree of freedom in maths lessons. They were usually given choices between different projects to work on and they were encouraged to decide the nature and direction of their work. Sometimes the different projects varied in difficulty and the teachers guided students towards projects that they thought were suited to their strengths. During one of my visits to the classrooms students were working on a project called '36 fences'. The teacher started the project by asking all of the students to gather round the board at the front of the room. There was a lot of shuffling of chairs as students made their way to the front, sitting in an arc around the board. Jim, the teacher, explained that a farmer had 36 individual fences, each of them 1 metre long, and that he wanted to put them together to make the biggest possible area. Jim then asked students what shapes they thought the fences could be arranged into. Students suggested a rectangle, triangle or square, "How about a pentagon?" Jim asked, the students thought about this and talked about it. Jim asked them whether they wanted to make irregular shapes allowable or not.

After some discussion Jim asked the students to go back to their desks and think about the biggest possible area that the fences could make. Students at Phoenix Park were allowed to choose who they worked with, and some of them left the discussion to work alone, while most worked in pairs or groups of their choosing. Some students began by investigating different sizes of rectangles and squares, some plotted graphs to investigate how areas changed with different side lengths. Susan was working alone, investigating hexagons, and she explained to me that she was working out the area of a regular hexagon by dividing it into six triangles and she had drawn one of the triangles separately. She explained that she knew that the angle at the top of each triangle must be 60 degrees, so she could draw the triangles exactly to scale using compasses and find the area by measuring the height.

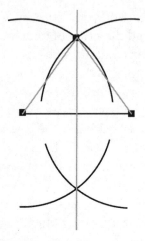

I left Susan working and moved to sit with a table of boys. Micky had found that the biggest area for a rectangle with perimeter 36 is 9 × 9 – this gave him the idea that shapes with equal sides may give bigger areas and he started to think about equilateral triangles. Micky seemed very interested by his work and he was about to draw an equilateral triangle when he was distracted by Ahmed who told him to "forget triangles!". Ahmed explained, excitedly, that he had found that the shape with the largest area made of 36 fences was a 36-sided shape. Ahmed told Micky to find the area of a 36-sided shape too and he leant across the table, explaining how to do this. He explained that you divide the 36-sided shape into triangles and all of the triangles must have a 1cm base, Micky joined in saying 'yes and their angles must be 10 degrees!' Ahmed said 'yes but you have to find the height and to do that you need the tan button on your calculator, T-A-N, I'll show you how, Mr Collins has just shown me'.

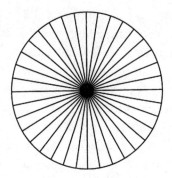

Micky and Ahmed moved closer together, using the tangent ratio to calculate the area.

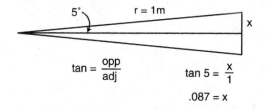

As the class worked on their investigations of 36 fences many of the students divided shapes into triangles. This gave the teacher the opportunity to introduce students to trigonometric ratios. The students were excited to learn about trig ratios as they enabled them to go further in their investigations and the methods were useful to them.

At Phoenix Park, the teachers taught mathematical methods to help students solve problems. Students learned about statistics and probability, for example, as they worked on a set of activities called 'Interpreting the World'. During that project they interpreted data on university attendance, pregnancies, football results and other issues of interest to them. Students learned about algebra as they investigated different patterns and represented them symbolically, they learned about trigonometry in the '36 fences' projects and by investigating the shadows of objects. The different projects were carefully chosen by the teachers to interest the students and to provide opportunities for learning important mathematical concepts and methods. Some projects were applied, requiring that students engage with real world situations, other activities started with a context, such as 36 fences, but led into abstract investigations. As students worked they learned new methods, they chose methods they knew and they adapted and applied both. Not surprisingly the Phoenix Park students came to view mathematical methods as flexible problem solving tools. When I interviewed Lindsey in the second year of the school she described the maths approach saying: *"Well if you find a rule or a method, you try and adapt it to other things. When we found this rule that worked with the circles we started to work out the percentages and then adapted it, so we just took it further and took different steps and tried to adapt it to new situations."*

Students were given lots of choices as they worked. They were allowed to choose whether they worked in groups, pairs or alone. They were often given choices about activities to work on and they were always encouraged to take problems in directions that were of interest to them and to work at appropriate levels. Most of the students liked this mathematical freedom, as Simon told me: *" You're able to explore, there's not many limits and that's more interesting."* Discipline was very relaxed at Phoenix Park and students were also given a lot of freedom to work or not work.

Amber Hill School.

At Amber Hill school the teachers used the traditional approach that is commonplace in England. The teachers began lessons by lecturing from the board, introducing students to mathematical methods, students then worked through exercises in their books. When the students at Amber Hill learned trigonometry,

Amber Hill's approach stood in stark contrast to Phoenix Park's – the Amber Hill students spent more time on task but they thought maths was a set of rules that needed to be memorized and few of them developed the levels of interest developed by the Phoenix Park students. In lessons the Amber Hill students were often successful, getting lots of their questions right in their exercises, but they often got them right, not by understanding the mathematical ideas but by following cues. For example, the biggest cue telling students how to answer a question was the method they had just had explained on the board. The students knew that if they used the method they had just been shown they were probably going to get the questions right. They also knew that when they moved from exercise A to exercise B they should do something slightly more complicated. Other cues they followed were – they knew that they should use all the lines given to them in a diagram and all the numbers in a question, if they didn't use them all they thought they were doing something wrong. Unfortunately the same cues were not present in the exams, as Gary told me, when describing to me why he found the exams hard: *"It's different, and like the way it's there like – not the same. It doesn't like tell you it, the story, the question; it's not the same as in the books, the way the teacher works it out."* Gary seemed to be suggesting, as I had seen in my observations, that the story or the question in their books often gave away what they had to do, but the exam questions didn't. Trevor also talked about cues when he explained why his exam grade hadn't been good: *"You can get a trigger, when she says like "simultaneous equations" and "graphs" or, "graphically". When they say like – and you know, it pushes that trigger, tells you what to do."* I asked him "What happens in the exam when you haven't got that?" he gave a clear answer: *"You panic."*

The GCSE examination in mathematics is a three-hour, traditional test made up of short mathematics questions. Despite the difference in the two school approaches, the students' preparation for the examination was fairly similar as both schools gave students past examination papers to work through and practice. At Phoenix Park the teachers stopped the project work a few weeks before the examination and focused upon teaching any standard methods that students may not have met. They spent more time lecturing from the board, and classrooms looked similar (briefly) to those at Amber Hill.

Many people expected the Amber Hill students to do well on the examinations, as their approach was meant to be examination oriented, but it was the Phoenix Park students who attained *significantly higher examination grades*. The Phoenix Park students also achieved higher grades than the national average, despite having started their school at significantly lower levels than national average. The examination success of the students at Phoenix Park surprised people in England and the research study was reported in most of the national newspapers. People believed that a project-based approach would result in great problem solvers, but they had not thought that an approach that was relaxed and project based without endless hours of practicing procedures, could also result in higher examination grades.

National Newspaper Headlines

The Independent	**Trendy teachers are top in maths** WENDY BERLINER — mixed ability groups in every maths lesson. There was very lit-
The Times	**Progressive methods 'best for maths'** Progressive maths teaching led to better results — by Diana Hofkins — approach, did not apply themselves, and would have preferred textbooks.
The Guardian	**They've got the right formula**

These were not quite the headlines I would have chosen but the approach was gaining rightful attention.

The Amber Hill students faced many problems in the examination, which they were not expecting as they had worked so hard in lessons. In class the Amber Hill students had always been shown methods and then practiced them. In the examination they needed to *choose* methods to use and many of them found that difficult. As Alan explained to me:*" It's stupid really 'cause when you're in the lesson, when you're doing work — even when it's hard — you get the odd one or two wrong, but most of them you get right and you think well when I go into the exam I'm gonna get most of them right, 'cause you get all your chapters right. But you don't."* Even in the examination questions when it was obvious which methods to use, the Amber Hill students would frequently confuse the steps they had learned. For example, when the Amber Hill students answered a question on simultaneous equations they attempted to use the standard procedure they had been taught, but only 26 per cent of students used the procedure correctly. The rest of the students used a confused and jumbled version of the procedure and received no marks for the question.

The Phoenix Park students had not met all of the methods they needed in the examination, but they had been taught to solve problems and they approached the examination questions in the same flexible way as they approached their projects – choosing, adapting and applying the methods they had learned. I asked Angus whether he thought there were things in the exam that they hadn't seen before. He thought for a while and said: *"Well, sometimes I suppose they put it in a way which throws you, but if there's stuff I actually haven't done before I'll try and make as much sense of it as I can, try and understand it and answer it as best as I can, and if it's wrong, it's wrong".*

The Phoenix Park students didn't only do better on the examinations. As part of my research I investigated the usefulness of the approach to students' lives. One way I measured this was by had experienced[5]. I had frequently been asked about the future of the students after they had left their schools and so I decided to find out. I sent surveys to the ex-students' addresses and followed up the surveys with interviews. As part of the survey I asked the young people what jobs they were doing. I then classified all the jobs and put them onto a scale of social class, which gives some indication of the professionalism of their jobs and the salaries they would have received. This showed something very interesting. When the students were in school their social class levels (determined from parents' jobs) had been equal. Eight years after my study the Phoenix Park young adults were working in more highly skilled or professional jobs than the Amber Hill adults, even though the school achievement range of those who had replied to the surveys from the two schools had been equal. Comparing the jobs of the children to their parents, 65% of the Phoenix Park adults had moved into jobs that were more professional, compared with 23% of Amber Hill adults. Fifty-two per cent of the Amber Hill adults were in less professional jobs than their parents, compared with only 15% of the Phoenix Park adults. At Phoenix Park there was a distinct upward trend in careers and economic well-being, at Amber Hill there was not despite the fact that Phoenix Park was in a less prosperous area.

In addition to the survey I conducted follow up interviews with the young adults. I contacted a representative group from each school, choosing people with comparable examination grades. In interviews the Phoenix Park adults communicated a positive approach to work and life, describing the ways they used the problem solving approach they had been taught in their mathematics classrooms to solve problems and make sense of mathematical situations in their lives. Adrian had attended university and studied economics and he told me that *"You often get lots of stuff where there will be graphs of economic situations in countries and stuff like that. And I would always look at those very critically. And I think the maths that I've learned is very useful for being able to actually see exactly how it's being presented, or whether it's being biased."* When I asked Paul, who was a senior regional hotel manager, whether he found the mathematics he had learned in school useful he said that he did, telling me: *"I suppose there was a lot of things I can relate back to maths in school. You know, it's about having a sort of concept, isn't it, of space and numbers and how you can relate that back. And then, okay, if you've got an idea about something and how you would then use maths to work that out. (. . .) I suppose maths is about problem-solving for me. It's about numbers, it's about problem-solving, it's about being logical."*

Whereas the Phoenix Park young adults talked of maths as a problem solving tool, and they were generally very positive about their school's approach, the Amber Hill students could not understand why their school's mathematics approach had prepared them so badly for the demands of the workplace. Bridget spoke sadly when she said: *"It was never related to real life, I don't feel. I don't feel it was. And I think it would have been a lot better if I could have seen what I could use this stuff for, and just basically . . . because then it helps you to know <u>why</u>. You learn <u>why</u> that is that, and <u>why</u> it ends up at that. And I think definitely relating it to real life is important."* Marcos was also puzzled as to why the school's maths approach had seemed so removed from the student's lives and work: *"It was something where you had to just remember in which order you did things, and that's it. It had no significance to me past that point at all—which is a shame. Because when you have parents like mine who keep on about maths and how important it is, and having that experience where it just seems to be not important to anything at all really. It was very abstract. And it was kind of almost purely theoretical. As with most things that are purely theoretical, without having some kind of association with anything tangible, you kind of forget it all."*

My book of the study of the approaches at Phoenix Park and Amber Hill – *Experiencing School Mathematics* – won a national book award in England and has been read by thousands of British, American and other readers. Many teachers have contacted me saying that they would like to teach through a problem solving approach similar to Phoenix Park's but they cannot because of lack of support from departments, schools and parents. I know that parents often feel powerless to help with the bad maths teaching that they hear about from their children, often thinking that it is just the way maths has to be – painful and irrelevant. But this is not the case and parents can be very powerful in bringing about change.

2. The Communicative Approach.

Railside High School is an inner-city high school in California and lessons are frequently interrupted by the sound of speeding trains. As with many inner-city schools the buildings look as though they are in need of some repair, but Railside is not like other inner-city schools. Advanced level maths classes are often badly attended or non-existent in many schools, but at Railside they are packed with eager and successful students. When I have taken visitors to the school and we have stepped inside the maths classrooms, they have been amazed to see all the students hard at work, engaged and excited about maths. In the United States, high schools teach children between the ages of 13 and 18 and mathematics is taught in courses – typically algebra, geometry, advanced algebra, pre-calculus and calculus.

In most high schools algebra and geometry are required before students can graduate. I first visited Railside in 1999, when I was a professor at Stanford University, because I had learned that the teachers collaborated and planned teaching ideas together and I was interested to see their lessons. I saw enough in that visit to invite the school to be part of a longitudinal research project investigating the effectiveness of different mathematics approaches. Some four years later, after we had followed 700 students through three high schools, observing, interviewing and assessing students, we knew that Railside's approach was both highly unusual and highly successful.

The mathematics teachers at Railside originally taught using traditional methods, but the teachers were unhappy with the high failure rates among students and the students' lack of interest in maths, so they worked together to design a new approach. Teachers met together over several summers to devise a new maths curriculum – first they worked on the school's algebra course and they later improved all the courses on offer. They also changed their classes from those taught in sets, to those taught in mixed attainment groups. In most algebra classes in the US students work through questions designed to give practice on mathematical techniques such as 'factoring polynomials', or 'solving inequalities'. At Railside the students learned the same methods but the curriculum was organized around bigger mathematical ideas, with unifying themes such as "What is a linear function?" A focus of the Railside approach was 'multiple representations' – the students learned about the different ways that mathematics could be communicated through words, diagrams, tables, symbols, objects, and graphs. As they worked the students would frequently be asked to explain work to each other, moving between different representations and communicative forms. When we interviewed students and asked them what they thought maths was, they did not tell us that it was a set of rules, as most students do, instead they told us that maths was a form of communication, or a language, as one young man explained: *"Math is like kind of a language, because it has got a whole bunch of different meanings to it, and I think it is communicating. When you know the solution to a problem, I mean that is kind of like communicating with your friends."*

In one of the lessons I observed students were learning about functions. The students had been given what the teachers referred to as "pile patterns". Different students had been given different patterns to work with. Pedro was given the pattern below, which includes the first 3 cases:

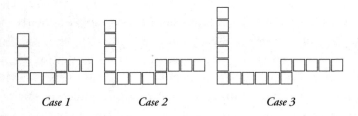

Case 1   Case 2   Case 3

The aim of the activity was for students to work out how the pattern was growing (you could try this too) and to represent this as an algebraic rule, a t-table, a graph and a generic pattern. Students also needed to show the 100th case in the sequence, having been given the first 3 cases.

Pedro started by working out the numbers that went with the first 3 cases and he put these in his t-table:

Case Number	Number of Tiles
1	10
2	13
3	16

He noted at this point that the pattern was "growing" by 3 each time. Next he tried to *see* how the pattern was growing in his shapes, and after a few minutes he saw it! He could see that each of the 3 sections grew by 1 each time. He represented the first two cases in this way:

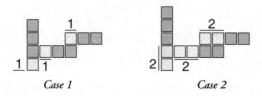

Case 1 *Case 2*

He could see that there were 7 tiles that always stayed the same and were present in the same positions (this was the way he visualized the pattern growth, but there are other ways of visualizing it). In addition to the 'constant' 7 there were tiles that grew with every case number. So, for example, if we just look at the vertical line of tiles:

We see that in case 1, there is 1 at the bottom, plus 3. In case 2 there is 2 plus 3. In case 3 there would be 3 plus 3 and

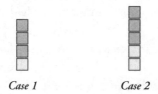

Case 1 *Case 2*

in case 4 there would be 4 plus 3 and so on. The three is a constant but there is one more added to the lower section of tiles each time. We can also see that the growing section is the same size as the case number each time. When the case is 1 the total number of tiles is 1 plus 3, when the case is 2 the total is 2 plus 3, we can assume from this that when it is the 100th case, there will be 100 + 3 tiles. This sort of work – considering, visualizing and describing patterns is at the heart of mathematics.

Pedro represented his pattern algebraically in the following way:

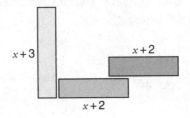

Where x stood for the case number. By adding together the three sections he could now represent the whole function as $3x + 7$.

At this point I should explain something about algebra, for those who find this example totally bewildering. When a friend of mine read about this pattern she was utterly lost and I realized that her confusion came from the way she had learned algebra in her traditional maths classes. She looked at the pattern with me and saw that the student had represented it as $3x + 7$ and she asked me, so what is x? I said that x was the case number, so in the first case x is 1, in the second x is 2 etc. This completely confused her and I realized that she was confused because to her x was always meant to be *a single* number. She had spent so many years of maths classes "solving for x" – rearranging equations to find out what number x was, that she, like millions of school children, had missed the most important point about algebra – that x is used to represent a *variable*. The reason that algebra is used so pervasively by mathematicians, scientists, medics, computer programmers and many other professionals is because patterns – that grow and change – are central to their work and to the world, and algebra is a key method in describing and representing them. My friend could see that the pattern increased by a different amount each time but was just not used to using algebra to represent a *changing* quantity. But the task in this problem – to find a way of visualizing and representing the pattern, using algebra to describe the changing parts of the pattern – is extremely important algebraic work. The way that most people learn algebra hides the meaning of algebra, it stops them from using it appropriately and it hinders their ability to see the usefulness of algebra as a problem-solving tool in mathematics and science.

Pedro was pleased with his work and he decided to check his algebraic expression with his t-table. Satisfied that $3x + 7$ worked, he set about plotting his values on a graph. I left the group as he was eagerly reaching for graph paper and coloured pencils. The next day in class I checked in with him again. He was sitting with 3 other boys and they were designing a poster to show their 4 functions. Their 4 desks were pushed together and covered by a large poster that was divided into 4 sections. From a distance the poster looked like a piece of mathematical artwork with colour-coded diagrams, arrows connecting different representations to each other and large algebraic symbols.

After a while the teacher came over and looked at the boys' work, talking with them about their diagrams, graphs, and algebraic expressions, probing their thinking to make sure they understood the algebraic relationships. He asked Pedro where the 7 (from $3x + 7$) was represented on his graph. Pedro showed the teacher and then decided to show the $+7$ in the same colour on his tile patterns, his graph and in his algebraic expression. The communication of key features of functions using colour-coding was something all students were taught in the Railside approach, to give meaning to the different representations. This helped the students learn something important – that the algebraic expression shows something tangible and that the relationships within the expression can also be seen in the tables, graphs and diagrams.

Juan, sitting at the same table, had been given a more complicated, non-linear pattern that he had colour coded in the following way:

See if you can work out how the pattern is growing and the algebraic expression that represents it!

As well as producing posters that showed linear and non-linear patterns, the students were asked to find and connect patterns, within their own pile patterns, and across all four teammates' patterns, and to show the patterns using technical writing tools. Because some of the pile patterns were non-linear, this was a complicated task for year 9 students, and it provoked much discussion, consternation, and learning! One of the aims of the lesson was to teach students to look for patterns within and among representations and to begin to understand generalization.

The tasks at Railside were designed to fit within the separate subject areas of algebra and geometry, in line with the tradition of content separation in American high schools. Still the problems the students worked on were open enough to be thought about in different ways and they often required that students represent their thinking using different mathematical representations, emphasizing the connections between algebra and geometry.

The Railside classrooms were all organized in groups and students helped each other as they worked. The Railside teachers paid a lot of attention to the ways the groups worked together and they taught students to respect the contributions of other students, regardless of their prior attainment or their status with other students. One unfortunate but common side effect of some classroom approaches is that students develop beliefs about the inferiority or superiority of different students. In the other classes we studied, that were taught traditionally, students talked about other students as smart and dumb, quick and slow. At Railside the students did not talk in these ways. This did not mean that they thought all students were the same, but they came to appreciate the diversity of the class and the various attributes that different students offered. As Zane described to me: *"Everybody in there is at a different level. But what makes the class good is that everybody's at different levels so everybody's constantly teaching each other and helping each other out."* The teachers at Railside followed an approach called 'complex instruction' (http://cgi.stanford.edu/group/pci/cgi-bin/site.cgi), which is a method designed to make group work more effective and to promote equity in classrooms. They emphasized that all children were "smart" and had strengths in different areas and that everyone had something important to offer when working on maths.

As part of our research project we compared the learning of the Railside students to that of a similar sized group of students in two other high schools, who were learning mathematics through a more typical, traditional approach. In the traditional classes the students sat in rows at individual desks, they did not discuss mathematics, they did not represent algebraic relationships in different ways and they generally did not work on problems that were applied or visual. Instead the students watched the teacher demonstrate procedures at the start of lessons and then worked through textbooks filled with short, procedural questions. The two schools using the traditional approach were more suburban and students started the schools with higher mathematics achievement levels than the students at Railside. But by the end of the first year of our research study the Railside students were achieving at the same levels as the students in the suburban schools on tests of algebra; by the end of the second year the Railside students were outperforming the other students on tests of algebra and geometry[6].

In addition to the high achievement at Railside, the students learned to enjoy and like maths. In surveys administered at various times during the 4 years of the study, the students at Railside were always significantly more positive and more interested in mathematics than the students from the other classes. By year 12, a staggering 41% of the Railside seniors were in advanced classes of pre-calculus and calculus, compared with only 23% of students from the traditional classes. Further, when we interviewed 105 students (mainly from year 12) at the end of the study, about their future plans almost all of the students from the traditional classes said that they had decided not to pursue mathematics as a subject – even when they had been successful. Only 5% of students from the traditional classes planned a future in mathematics compared with 39% of Railside students.

There were many reasons for the success of the Railside students. Importantly, the students were given opportunities to work on interesting problems that required them to think, and not just reproduce methods, and they were required to discuss mathematics with each other, increasing their interest and enjoyment. But there was another important aspect of the school's approach that is much more rare – the teachers enacted an expanded conception of mathematics and "smartness". The teachers at Railside knew that being good at mathematics involves many different ways of working, as mathematicians' accounts tell us. It involves asking questions, drawing pictures and graphs, rephrasing problems, justifying methods and representing ideas, in addition to calculating with procedures. Instead of just rewarding the correct use of procedures the teachers encouraged and rewarded all of these different ways of being mathematical. In interviews with students from both the traditional and the Railside classes, we asked students what it took to be successful in maths class. Students from the traditional classes were unanimous: they all said that it involved paying careful attention – watching what the teacher did and then doing the same. When we asked students from the Railside classes they talked of many different activities including: asking good questions, rephrasing problems, explaining ideas, being logical, justifying methods, representing ideas and bringing a different perspective to a problem. Put simply, because there were many more ways to be successful at Railside, many more students *were* successful.

Janet, one of the year 9 students, described to me the way that Railside was different from her middle school experience saying that: *"back in middle school the only thing you worked on was your math skills. But here you work socially and you also try to learn to help people and get help. Like you improve on your social skills, math skills and logic skills."* Jasmine also talked about the variety in Railside's approach saying that *"With math you have to interact with everybody and talk to them and answer their questions. You can't be just like "oh here's the book, look at the numbers and figure it out."* We asked Jasmine why maths was like that, she answered: *"It's not just one way to do it (. . .) It's more interpretive. It's not just one answer. There's more than one way to get it. And then it's like: "why does it work"?* The students highlighted the different ways that mathematics problems could be solved and the important role played by mathematical justification and reasoning. The students at Railside recognized that helping, interpreting and justifying were critical and valued in their maths classes.

In addition the teachers at Railside were very careful about identifying and talking to students about all the ways they were "smart". The teachers knew that students – and adults – were often severely hampered in their mathematical work by thinking they were not smart enough. They also knew that all students could contribute a great deal to mathematics and so they took it upon themselves to identify and encourage students' strengths. This paid off and any visitor to the school would have been impressed by the motivated and eager students who believed in themselves and who knew they could be successful in mathematics.

The two teaching approaches I have described were the subject of comprehensive research studies and although they were conducted in different countries, the findings pointed to the same conclusion: Students need to be actively involved in their learning and they need to be engaged in a broad form of mathematics, using and applying methods, representing and communicating ideas. In both of the successful approaches I have reviewed the teachers did not decide what students could do, before they taught them, and they chose to group children in mixed attainment groups. They then gave students mathematical activities that were broad and that allowed different students to take them to different levels. During these times they encouraged all students to believe in themselves and to feel that they were good at maths. Because the activities were broad it meant that everybody found parts that they were good at and that made them feel good about maths. The breadth of the projects also meant that they valued all of the different ways of being mathematical, and students were not simply repeating procedures. The students were encouraged to talk about maths, to ask questions, to communicate maths in different ways, to decide on the appropriateness of different methods and to adapt methods in order to solve problems. The active approaches and the breadth of maths in which they engaged meant that students enjoyed maths, many of them were successful at maths, and there was gender equity in both schools. I will explain in chapter 9 how parents can support teachers in moving to approaches that could be as effective as those in Phoenix Park and Railside schools.

4 / Banishing the Monsters

Moving to More Effective Forms of Assessment.

On Monday 12th May, 2008, BBC Panorama aired a programme entitled 'Testing to Destruction'. It was a programme showing the negative and destructive effects of the testing system we now have in English schools. During the programme researchers asked primary aged children to draw the creature they thought of when thinking about SATs[1]. The children responded with drawings of horrific monsters. As one child explained of his monster creation "he rips your body in half and eats your guts". These are the images that fill young children's minds because of the pressure they face by the tests now used in schools.

England used to have an assessment system that was the envy of the world. Teachers assessed students throughout school, giving parents regular reports of progress. The first high-stakes test that the students would encounter was the GCSE[2] when they were 16 years old. Prior to that time children were encouraged to *learn* in classrooms, and they did not waste months of school time being coached to perform on tests whose sole purpose was to compare schools. Teachers, for the most part, used to concentrate on the most important and worthwhile subjects and address the breadth of the curriculum. All of that has changed as English children become the most over-tested children in the world[3].

The 1988 Education Reform Act, put into place by Margaret Thatcher's government, marked a new era in schools with a defined curriculum and compulsory tests for 7, 11 and 14 year olds. In Northern Ireland, Scotland and Wales, SATs for 7 year olds have been abolished, as politicians realised that 7 year-old children are much too young to be put under the pressure that such tests incur, in England they continue. When students are aged 11 they take year 6 SATs and the results of these are used to judge schools with schools being placed into league tables. The consequences of such crude reporting are entirely negative and the year 6 league tables have now been abolished in Northern Ireland, Scotland and Wales, but not in England. We must now wonder why it is that England is out of synch with the rest of the UK, especially when the other UK countries abolished age 7 SATs and age 11 SAT league tables, for very good reasons.

When students enter year 6 in English primary schools they are entering a year of narrow test preparation. In 2003 the government agency responsible for curriculum and assessment (the Qualifications and Curriculum Authority, or QCA) asked schools how long they spent preparing children for the year 6 SATs. They found that the average primary school spends a staggering ten hours per week in test preparation in the period between Christmas and May. In the four-month build up to test time schools spend 40% of their time coaching children[4]. This preparation means that students are no longer learning the breadth of the curriculum, instead they are forced to practice past-papers, and many become extremely de-motivated in the tested subjects – especially maths. Most important of all the young children are put under extreme pressure to perform, resulting in high levels of stress and perceived failure for many students, when they do not make the grade that they are pushed towards. This stress was illustrated poignantly by a 10 year-old student telling interviewers: *"I'm really scared about the SATs. Mrs O'Brien came and talked to us about our spelling and I'm no good at spelling and David (the class teacher) is giving us times tables tests every morning and I'm hopeless at times tables so I'm afraid I'll do the SATs and I'll be a nothing"*. When the interviewer tried to convince her that she could never be a "nothing", even if zero was her recorded SAT-level, the young girl insisted that the tests would make her one. The researchers reported that even though she was "an accomplished writer, a gifted dancer and artist and good at problem solving" the tests made her feel as though she was an "academic non-person".[5]

The irony of this situation is that the year 6 SAT gives no educational benefit to children – its sole purpose is to compare schools. Warwick Mansell has written a very clear and readable account of the impact of our country's testing system[6]. He points out that primary schools cannot even use the year 6 test results to help children, acting on their strengths and weaknesses, because they receive the results as the children are about to leave primary school. Secondary schools do not value the results of the year 6 SATs as they know that students have been coached narrowly – the SATs serve one purpose only – which is to compare schools. Students go through enormous stress and lose about half a year of their education, for a test that gives them no benefit whatsoever.

The problem with the situation we have is not that national assessments are used in schools, but that they are used to compare schools in league tables. When schools are ranked in this way and when teachers' and head-teachers' jobs are judged by such narrow measures, teachers respond by coaching children and putting them under pressure. League tables based on SAT results are also a crude and unsophisticated way of comparing schools. Children who come from privileged homes do better on tests – that fact alone is enough to cast doubt on the methods now used to judge schools. Schools used to be evaluated by teams of OFSTED inspectors who spent time in the schools watching lessons, talking to children and teachers, and developing careful and nuanced judgements that included consideration of test results, but were not dominated by them. Now OFSTED inspectors spend less time in schools because they judge the schools primarily on their test scores. In 2003 members of the National Association of Head Teachers called for a ban on the publication of test results saying that they amounted to 'a crime against children'[7] but still the system continues. We need assessments in schools and we need schools to be inspected and judged according to high standards but England has now reached a ridiculous situation and children face being, 'tested to destruction'. (Panorama).

Warwick Mansell, and many other analysts, point out that we are in a time of 'hyper-accountability' as schools, teachers and head-teachers are constantly compared to each other using test scores, with punitive measures brought against low-performing schools. This system needs to change, with assessments being used to help children improve their learning and to give good information to parents, rather than to compare and order schools. Test results can be given to OFSTED inspectors for consideration, but they should not be used to rank schools. Instead of our culture of hyper-accountability, based upon narrow measures, schools should be encouraged to develop a culture of self-evaluation, monitoring students' learning using broad and valid assessments. No school lacks motivation to improve the learning experiences of children. In a system of self-evaluation children would still receive regular assessments and reports to parents. The critical difference between the two systems is that a culture of self-evaluation allows schools to focus upon the most important learning experiences for children, and give regular and good assessments, whereas a hyper-accountability culture means that schools subject children to months of coaching for external tests, and students are put under significant amounts of pressure as they are prepared for high-stakes testing and reporting.

Research that has investigated the impact of high-stakes testing in the US, where they have had such testing for many years, has shown its effects to be almost entirely negative. High stakes tests have now been implemented across the United States but in the days when they were only used in some states researchers were able to conduct comparative studies. Audrey Amrein and David Berliner,[8] for example, considered the impact of high stakes tests by comparing states that did and did not use them. They examined the effect of the tests by asking whether they improved students' learning. They didn't just examine learning by looking to see if scores on the tests themselves improved – as that can be done by teaching to the test and excluding students – they measured learning using different assessments, such as the well-regarded NAEP assessments and the AP examinations[9]. They found that in 17 of the 18 states that used high stakes testing student learning remained *at the same level or went down*. The high-stakes tests also produced a number of unintended negative consequences, such as teachers leaving the profession, increased school drop out rates, and cheating by teachers and schools.

In addition to improving the ways that assessments are used in England, we need to improve the nature of the assessments used. Tests drive the curriculum and determine the content goals that teachers work towards. It is therefore critically important that mathematics tests and exams assess *the most important mathematics*. In our present system the Mathematics Associations of England unite in their concern that SATs and GCSE's are too narrow and do not assess the most important areas of mathematics. In addition, the extensive time children face being coached for tests means that some very important mathematical acts are neglected. The Mathematical Association, for example, argues that the current assessment system encourages schools to focus on test preparation which means 'the exclusion of more interesting and challenging problems at all levels. These are the very things that are of importance to employers and higher education, because they stimulate interest and encourage independent thinking'.[10] OFSTED inspectors agree, in preparing a report for a critical national review of mathematics,[11] inspectors reported that the best teaching they saw 'gave a strong sense of the coherence of mathematical ideas; it focused on understanding mathematical concepts and developed critical thinking and reasoning' but that much of the teaching they saw 'presented mathematics as a collection of arbitrary rules and provided a narrow range of learning activities (that) did not motivate students and limited their achievement'. The reason that the OFSTED inspectors gave for the narrow version of mathematics they saw was schools 'focusing heavily on examination questions' which 'enabled students to pass examinations, but did not necessarily enable them to apply their knowledge independently in different contexts'[12]. The narrow versions of mathematics that are assessed in SATs and GCSE's, is a critical issue for the future of our country.

The narrow assessments used in schools are the main reason that students have narrow mathematical experiences in schools and they experience teaching that OFSTED describe as de-motivating. Alfie Kohn quotes from a teacher who had to stop one of her most successful teaching activities because of the tests she was forced to prepare students for. She used to ask her middle school students to become an expert in a subject, developing their researching and writing skills. This was an experience that her students remembered for years, that they looked back on as a highlight of school, but that she was forced to end. As Kohn writes: '"the most engaging questions kids bring up spontaneously – 'teachable moments' – become annoyances". Excitement about learning pulls in one direction; covering the material that will be on the test pulls in the other'.[13] The elimination of powerful learning experiences because they cannot be reduced to testable knowledge is having a serious effect in England. If OFSTED inspectors recognize that mathematics assessments are too narrow, and the Mathematical Associations[14] have published national reports expressing their concern for the narrow assessments, why do they continue?

Alongside the bleak picture of hyper-accountability that has invaded children's lives there is a glimmer of light on the assessment scene. The glimmer is a new approach to assessment that holds great potential for improving children's learning. The approach, called 'assessment for learning' (often shortened to A4L) is based upon important principles concerning knowledge and self-awareness and it is an approach that could be used by parents when working with children, to great effect. For the approach to work it needs to be implemented in a way that is true to its founding principles. This has happened in Scotland, resulting in widespread improvements[15], but the implementation of A4L in England has been much less careful, with varied results. In this chapter I will explain the key principles, and the methods that teachers and parents can use, as they hold the power to increasing children's experiences and understandings, significantly.

Assessment For Learning

'Assessment for learning', is an approach that was brought to public attention by Professors Paul Black and Dylan Wiliam in 1998. They showed that if A4L was implemented well it would have an impact on learning so great that it would raise a countries ranking in international comparisons from the middle of the pack to a place in the top 5[16]. The methods are now used throughout the world, although not always in the ways that were intended[17]. A4L is based upon the principle that students should have a full and clear sense of what they are learning, of where they are in the path towards mastery, and what they have to do to become successful. Students are given the knowledge and tools to become self–regulatory learners, so that they are not dependent upon following somebody else's plans, with little awareness of where they are going, or what they might be doing wrong.

It may seem obvious that learners should be clear about what they are learning and what they need to do to be successful but in most mathematics classrooms the students have very little idea. I have visited hundreds of classrooms and stopped at students' desks to ask them what they are working on. In traditional classrooms students usually tell me what page they are on, or what exercise they are working through. If I ask them, 'but what are you actually doing?' they say things like "oh I'm doing number 3". Students are usually able to tell you the titles of chapters they are working on but they really do not have a clear sense of the mathematical goals they are pursuing, the ways that the exercises they work through are linked to the bigger goals they are pursuing, or the differences between more and less important ideas. This makes it very difficult for students, or parents, to do anything to improve children's learning. Mary Alice White, a professor of psychology from Columbia University, has likened the situation to workers on a ship who may be given small tasks to do and complete them each day without having any idea of where the ship is heading, or the voyage they are undertaking.

The first part of the assessment for learning approach involves communication about what is being learned and where students are going. The second part involves making individual students aware of where they are in the path to success and the third part involves giving them clear advice about *how* to become more successful. The approach is called 'assessment *for* learning' rather than assessment *of* learning because it is designed to promote learning and all the information that is gained from assessment is made helpful to individual learners to propel them to greater levels of success.

So how is this achieved, and how does assessment for learning look different from traditional assessment, in the classroom? First of all, students are made aware of what they are, should and could be learning through a process of self and peer assessment. Teachers set out mathematical goals for students, not a list of chapter titles or tables of content, but details of the important ideas and the ways they are linked. For example, students might be given a range of statements that describe the understanding they should have developed during a piece of work, with statements such as: 'I have understood the difference between mean and median and know when each should be used'. The statements are clear for students to understand, and they communicate to them what they should be understanding from a piece of work. Students then assess their own, or their peers' work against the statements. By assessing work against statements and deciding what they have understood, students come to understand the goals of lessons much more clearly. In reviewing the goals of a lesson, week or unit of work, students start to become aware of what they should be learning and what the big ideas are. Parents can also review the criteria and become more knowledgeable about the ideas and knowledge students are working towards.

In studies of self-assessment in action researchers have found that students are incredibly perceptive about their own learning, and they do not over or under estimate it. They carefully consider goals and decide where they are and what they do and do not understand. In peer assessment students are asked to judge each other's work, again evaluating the work against clear criteria. This has been shown to be very effective, in part because students are often better able to hear criticism from their peers than from a teacher, and peers usually communicate in easily understood ways. It is also an excellent opportunity for students to become aware of the criteria against which they too are being judged. One way of managing peer assessment is to ask students to identify "two stars and a wish" – they select two things done well and one area to improve in their peers' work. When students are frequently asked to consider the goals of their learning, for themselves or their peers, they become very knowledgeable about what they are meant to be learning, and this makes a huge difference.

Evidence of the power of making students aware of what they are meant to be learning came from a very careful study conducted by two psychologists, Barbara White and John Frederiksen[18]. They worked with 12 classes of 30 students learning physics. Each class was taught a unit on force and motion with students in each class divided into an experimental or a control group. The control group used some periods of time each lesson discussing the work, while the experimental group spent the same time engaging in self and peer assessment, considering assessment criteria. The results were dramatic, with the experimental group outperforming the control group on three different assessments. The greatest gains were made by the students who had previously been the lowest achieving students. After the students spent time considering criteria and assessing themselves and their peers against them, the previously low achievers began to achieve at the same levels as the highest achievers. Indeed the year 8 students who reflected on the criteria scored at higher levels than year 12 students on tests of advanced physics[19]. White and Frederiksen concluded that the students were previously unsuccessful not because they lacked ability but because *they had not really known what they were meant to be focusing upon.*

When students are required to be aware of what they are learning and to consider whether they understand it or not this also gives important information to the teacher. For example, consider an A4L activity called "traffic lighting". In some versions of this students are asked to put a red, orange or green sticker on their work to say whether they understand new work well, a little or not at all. In other versions the teachers give students three paper cups – one green, one yellow and one red. If a student feels the lesson is going too fast they show the yellow cup, if they need the teacher to stop they show the red cup. At first researchers found that students were reluctant to show a red cup, but when teachers asked another student who was showing a green cup to offer an explanation, students became more willing to show a red cup when they were not understanding something. Other teachers used the traffic lighting to group students, sometimes having the greens and yellows work together to deal with problems between themselves, while the red pupils were helped by the teacher to deal with deeper problems. Most importantly the students themselves were being asked to think about what they knew and could do and what they needed more help on. This helps students and teachers tremendously as teachers gets feedback on their teaching in real time, rather than at the end of a unit or piece of work when it is too late to do anything about it, and they are able to deliver the most helpful information to students at an appropriate pace.

Methods of self and peer assessment serve the purpose of both teaching students about the goals of the work and the nature of high quality work and giving them information on their own understanding. For teachers these methods also give critical information on student understanding that can help them assist individuals in the best possible way and improve their own teaching. But, as Black and Wiliam have noted, these new methods require fundamental changes in the behaviour of both students and teachers. Students need to move from being passive learners to being active learners, taking responsibility for their own progress, and teachers need to be willing to lose some of the control over what is happening, which some teachers have described as scary but ultimately liberating. As one of the teachers in England reflected: 'What formative assessment has done for me is made me focus less on myself but more on the children. I have had the confidence to empower the students to take it forward.' (Robert, Two Bishop's School)[20]

The third part of the assessment for learning approach, after making students aware of what needs to be learned and how they are doing, involves helping students know *how* to improve, which is best achieved through diagnostic feedback. Psychology professors Maria Elawar and Lyn Corno[21] trained 18 year 6 teachers in three schools in Venezuela to give constructive written feedback in response to mathematics homework, instead of the scores they normally gave. The teachers learned to comment on errors, giving specific suggestions about how to improve, and to give at least one positive remark about the students' work. In an experimental study half the students received homework grades as normal, just receiving scores, and half received constructive feedback. The students receiving the constructive feedback learned *twice as fast* as the control-group students, the achievement gap between male and female students was reduced, and students' attitudes towards mathematics became significantly more positive.

In another interesting study, Ruth Butler, a professor of education at the Hebrew University of Jerusalem[22] compared three ways of giving feedback to different groups of students. One group received grades, one group received comments on their work, saying whether carefully explained criteria were matched or not, and one group received both grades and comments. What the researchers found was that the group receiving comments increased their performance significantly whereas those who received grades did not. More surprisingly, perhaps, those who received grades *and* comments did as poorly as those who only received grades. It turned out that those who received both grades and comments only focused upon the grades they received which served to override any comments. Diagnostic, comment-based feedback is now known to promote learning, and it should be the standard way in which students' progress is reported. Grades may be useful for communicating where students are in relation to each other, and it is fine to give them at the end of a term or year, but if they are given more frequently than that they will reduce the achievement of many. More typical reporting should be made up of feedback on the mathematics that students have learned, clear insights into the mathematics students are working towards, and advice on how to improve.

Assessment experts put it simply – "feedback to learners should focus on what they need to do to improve, rather than on how well they have done, and should avoid comparison with others".[23] This makes perfect sense – a coach tells athletes how to be better, they don't just tell them a grade, why should teachers just say, in so many words, you are a low achiever? As Royce Sadler, a professor of higher education, says: 'The indispensable conditions for improvement are that the student comes to hold a concept of quality roughly similar to that held by the teacher, is continuously able to monitor the quality of what is being produced during the act of production itself, and has a repertoire of alternative moves or strategies from which to draw at any given point.'[24] The beauty of the 'assessment for learning' approach is that it not only gives critical information to learners, but that it gives the same important information to teachers, helping them gear their teaching to their students' needs.

The main aim of 'assessment for learning' is the improvement of *classroom* assessment, but the methods have also been used to improve wide-scale assessments such as those used at national levels. This is important because we know that teachers gear their teaching to the national assessments that students take. In Queensland, Australia, for example, the government recognized the need for good assessments that could inform and improve learning rather than traditional assessments that simply ranked students, and they designed a system to ensure objectivity. The system was instigated in 1971 and has continued to evolve and be improved upon since then. It involves students taking two assessments, a school based 'course of work' in which teachers assess their students' work in class, which is moderated by a moderation committee; and a test of core skills. The test is used for the purpose of comparing performance across subjects, if there is a discrepancy between performance on the school work and the skills test then the former takes precedence. Importantly the school based assessment offers opportunity for good assessment tasks, criteria that students can consider as they work and diagnostic teacher feedback, combined with good, reliable information for parents and others. Other countries are now developing large-scale assessments that encourage learning but that can also be used to give unbiased, objective measurements of a students' learning through well developed systems to ensure that graders are following the same standards.

Analysts have found that spending money on assessment for learning is an extremely good investment. Dylan Wiliam compared the cost of investing in assessment for learning with two other innovations – increasing teachers' knowledge of mathematics, and reducing class size, both believed to be important in raising standards. He found that initiatives to increase teacher knowledge or reduce class size are expensive and promise small increases in learning. In contrast, training teachers to use 'assessment for learning' is a relatively cheap intervention, and it has been found to *double* the speed of learning. After teachers are trained in assessment for learning methods, at a cost of less than £2500 per teacher, students learn in six months what they would normally take a year to learn.

Good assessments are critical, whether they are national exams or classroom assessments. They should help students know what they are learning, provide students with the opportunity to show their understanding thoroughly, show students where they are in their learning, and when they are used in classrooms they should include feedback on how to improve. Importantly, students should be given information that refers to the content area being assessed, not to other students. Assessment for learning transforms students from passive receivers of knowledge to active learners who regulate their own progress and knowledge and propel themselves to higher levels of understanding. It also broadens teachers' assessment strategies, encouraging them to focus less on simple tests and more on the broad ways in which they can monitor student learning, including class work and student discussions and presentations. In chapter 9 I will outline some ways that parents may make use of the assessment for learning approach in work with their own children.

The implementation of A4L in England has not been careful, as schools were simply told to use the methods, rather than being given opportunities to learn them and engage with them. As a result some schools have focused upon some of the methods (such as versions of traffic lighting) without making the more fundamental shifts that are needed in the ways that they communicate learning goals and progress to students. Fundamentally A4L is an approach to 'assessment, teaching and learning'[25] and cannot be perceived as a bolt-on extra, a mere method of assessment, to be used once schools have decided how they are going to teach mathematics. Schools can only claim that they are using A4L if they engage in the three key activities of: communicating what is being learned and where students are going, helping individual students become aware of where they are in the path to success, and giving clear and diagnostic advice about how to become more successful.

An important obstacle to the success of A4L in England is that schools have been asked to implement A4L methods within a culture of external testing and evaluation. The two systems are fundamentally in conflict and it is difficult to encourage schools to use diagnostic assessments and feedback when they also have to prepare students for narrow tests and crude reporting. Nevertheless the fact that assessment for learning is written into policy for schools in England and Wales, and teachers have engaged with it enthusiastically, is extremely positive and what Paul Black has described as 'a bright star in an otherwise darker firmament'.[26]

In the future, I would expect to see a much more positive version of mathematics teaching in classrooms if SATs were abolished and students did not meet a high-stakes, high-pressure assessment until they were 16. Instead of SATs national assessments could be available for schools to use at the end of particular year groups. The assessments would come with data showing levels of good performance so that schools could get an indication of the relative position of their school students. Crucially, such data would be for school use and improvement of student learning, and for communicating with parents, not for drawing up league tables to be published in newspapers. The national assessments would need to be of a high quality, giving students opportunities to reason mathematically, and to problem-solve. This would mean that they would include some longer questions that students took time to complete and which would be graded by teachers against particular criteria and mark-schemes, developed with the main mathematical associations in

England[27]. As well as national assessments students would be regularly assessed using A4L methods in classrooms and they would receive diagnostic feedback, communicating how they may improve, that could be given to students and parents. Teachers and students could then focus on important mathematics, with schools preparing the sort of mathematical thinkers and problem solvers that our society needs. Students, armed with helpful, diagnostic advice would be enabled to reach much higher mathematical levels as they rid themselves of public labels and instead develop self-knowledge of where they are in mathematics, the targets they are aiming towards, and what they need to do to reach them.

5 / Making 'Low Ability' Children.

How Different Forms of Grouping Can Make or Break Children

In England we do something very cruel to children in maths classrooms, that sets us apart from just about every other country in the world. We tell children, from a very young age, that they are no good at maths. Other countries, particularly those who score very highly in maths assessments, do the opposite and they go to great lengths to communicate the idea that all children can do well in maths. Not only do they tell children that they can do well in maths, they *expect* children to do well in maths and they use teaching methods that *make sure* children do well in maths. In England we do not.

The idea that only some children can do well in maths is deeply cultural and it exists in England and the United States. It is also extremely harmful to children's learning and it is something that we need to change. Every child can do as well in maths at school as they do in other subjects – if they receive good teaching and people believe in them. The ability to do mathematics, at school levels, is not some sort of special gift that is bestowed upon a small number of children. Keith Devlin, mathematician and author, argues that being good at maths is simply part of being human, just the same as being human means we are good at speaking our native language. The gene for language development and mathematical development is the same[1]. Yet for some reason we all expect to speak English but we do not all expect to develop mathematical expertise.

The most worrying sign that our society believes that only some people can do maths, is the way that schools group children for maths. When we look at the grouping systems that are used in maths classrooms around the world, we see that England is very unusual as we have more ability grouping, with more divisions, applied at a much younger age than anywhere in the world. When my friend's daughter started her primary school last year she was told that children would be put into "sets" for maths when they were 5 years old. When she went to a "maths evening" at the school and some parents questioned the idea of telling children they were in a low maths group, the mathematics coordinator told the gathered parents that it would be fine, that children like it and that "they are happy to know their place in life". This judgment of children, with some of them being condemned to a life of low achievement, from the work they produce at ages 4 and 5, is nothing short of incredible, particularly when it goes against all research on children's learning and effective forms of grouping.

The latest statistics on ability grouping tell us that around 28% of primary schools use ability grouping for maths, in some year groups[2], 83% of schools use it in the Key Stage 3 years (ages 11-14) and 100% use it in Key Stage 4 (14-16)[3]. It is not unreasonable to think that children should be taught in similar achievement groups at some stage in their education, the question is when? If schools decide to group students too early, before anyone can know how well children *could* achieve, then their achievement will suffer. This makes the timing of grouping decisions absolutely critical as the most damaging message you can ever give a child is that they are "low ability", particularly when it comes to maths.

Research on the impact of ability grouping is very helpful in offering us a broader perspective on this important issue. When we consider the impact of ability grouping in the primary years we have to limit ourselves to research from England as other countries do not group students by "ability" at such a young age. In 2008 the government commissioned a review of research on grouping strategies in primary schools[4], as part of The Primary Review[5]. This considered a range of studies that have been conducted on mathematics grouping since the 1990's. The authors of the review drew unequivocal conclusions: 'The adoption of structured ability groupings has no positive effects on attainment but has detrimental affects on the social and personal outcomes for some children'[6]. The research that the reviewers consulted showed that ability grouping in primary schools had no academic benefits and severe negative consequences for children's development.

The researchers conducting the review realized that primary teachers choose to group children according to notions of 'ability' because they think that they can offer more appropriate work for children when they are in such groups. However the review found that 'the allocation of pupils to groups is a somewhat arbitrary affair and often depends on factors not related to attainment', also that although teachers think they are giving children in low groups more suitable work, 'the evidence suggests that many pupils find the work they are given is *inappropriate*; and often it is too easy.'[7] (2008, p27-28).

The evidence on this issue is clear – deciding that primary age children have "low ability" and grouping them in a low set, is damaging. Children develop at different times and rates in primary school and it is entirely possible for a child to struggle with maths in primary school but excel in later years (unless they are put into a low set in primary school). The least helpful action that any school can take with a young learner of maths is to group them with other low achievers, give them lower level work, have low expectations for them, and communicate to them that they are of "low ability".

At regular intervals researchers across the world collect data on the performance of children in different countries and consider the national factors that may impact performance. In the Third International Mathematics and Science Study (TIMSS), the highest achieving country, of 38 countries studied, was Korea, which was also the country with the least ability grouping. The high achievement of countries that do not use ability grouping and the low achievement of those that do was also a finding of the Second International Mathematics and Science Study (SIMSS,). This led analysts to conclude that countries who leave grouping to the *latest* possible moment or who use the *least* amount of grouping by ability are those with the highest achievement.

In England, the sensible time to group children with others of similar achievement is when they enter Key Stage 4 and they start preparing for examinations. At this time it makes sense for children who are taking the higher GCSE paper to work together, and those taking the foundation GCSE paper to work together, but research consistently shows that the years before that should be years when all students are encouraged to work at the highest possible mathematical level. When children begin secondary schools and they enter Key Stage 3, they are entering a new and very different phase of their educational career. It makes no sense to decide upon their future achievement based upon their work in primary school – all children should be prepared for high-level work. Yet 83% of secondary schools place children into ability groups for maths at this time.

What's more, whereas other countries that group children at a similar age, such as the United States (which also ranks at a low level in maths) typically place children into one of two groups at this time, schools in England use as many as ten sets for maths, giving *most* of the students in the year group a low number, that children associate with their ability. This harsh and extreme form of grouping, that is commonplace in English schools, shocks people from other countries in the world. Recently, in a UNICEF study of 20 countries, children in the UK were found to have the lowest self-esteem of all the children[8]. This fact cannot be separated from the harsh way we treat children in maths classrooms and the messages we send them about their capability. The situation is particularly extreme in mathematics, both because of the ridiculous forms of grouping used in maths (and in many schools maths departments are the only departments that use ability grouping) and because so many people believe that maths is a sign of general intelligence. When children are put into their low maths set, most of them simply believe that they have low ability.

Some have argued that the high degree of ability grouping used in England is a reflection of our desire to find and focus upon the high achieving students. But such an approach has some serious flaws, including the difficulty of identifying students correctly when children develop at different rates, and the creation of highly unequal schooling, which is one of the results identified by the international studies. In Japan, by contrast, (a country that consistently ranks highly in international comparisons) the main priority is to promote high achievement for all and teachers refrain from prejudging achievement, instead providing all students with complex problems that they can take to high levels. Japanese educators are bemused by the Western goal of sorting students into high and low 'abilities', as George Bracey, a professor from the George Mason University noted:

'In Japan there is strong consensus that children should not be subjected to measuring of capabilities or aptitudes and subsequent remediation or acceleration during the nine years of compulsory education. In addition to seeing the practice as inherently unequal, Japanese parents and teachers worried that ability grouping would have a strong negative impact on children's self-image, socialization patterns, and academic competition.'[9]

Lisa Yiu[10] was a student of mine while I was a professor at Stanford University. She was interested in the very different ways Japan and the US treat the grouping of students. She visited Japan and interviewed some of the mathematics teachers she observed, who explained to her why they do not use ability grouping:

'In Japan what is important is balance. Everyone can do everything, we think that is a good thing. (. . .) So we can't divide by ability.'

'Japanese education emphasizes group education, not individual education. Because we want everyone to improve, promote and achieve goals together, rather than individually. That's why we want students to help each other, to learn from each other (. . .), to get along and grow together – mentally, physically and intellectually.'

The Japanese approach of teaching students *"to help each other, to learn from each other, to get along and grow together – mentally, physically and intellectually"* is part of the reason for their high achievement. Research tells us that approaches that keep students as equal as possible and that do not group by 'ability' not only help those who would otherwise be placed in low sets, which seems obvious, but those who would be placed in high sets too.

In neither of my two longitudinal studies of students working in different mathematics approaches did I intend to research the impact of ability grouping, but it emerged as a powerful factor in both of them. In both studies the two successful schools were those that chose against ability grouping and used a different system of grouping, a result that mirrors the larger scale international findings. In my study in England I followed students in sets across the range from set 1 (the highest) to set 8 (the lowest). It did not surprise many people that students who were put into low sets achieved at low levels, partly because of the low level work they were given and partly because students gave up when they knew they had been put into a low set and labeled a low achiever. This was true of students from set 2 downwards. What was more surprising to people was that many of the students in set 1, the highest group, were also disadvantaged by the grouping. Students in set 1 reported that they felt too much pressure from being in the "top set". They felt the classes were too fast, they were unable to admit to not following or understanding and many of them started to dread and hate maths lessons. The students in the highest group should have felt good about their understanding of maths but instead they felt pressured and inadequate. After three years of ability-grouped classes the students achieved at significantly lower levels than students who had been grouped in mixed ability classes.

In the United States the ability grouping that is used is nowhere near as harsh as that in England. At around 7th and 8th grade students in the US typically get placed into one of two different level classes, which nevertheless determine their future for many years to come. In my 5-year US study we again followed students in schools that used ability grouping (called 'tracking' in the US) and one school (Railside) that did not. As I described in chapter 3, the results at Railside were outstanding, with students achieving at significantly higher levels. In addition, 47% of the senior Railside students took calculus and pre-calculus classes (similar to A-level), compared to 28% of students in the schools that used tracking. Interestingly, the students who were most advantaged by the mixed ability grouping at Railside were the initial high achieving students, who achieved at higher levels than students who were placed into high tracks in the other schools, and who improved their achievement more than any other students in their school.

As educators become more aware of the disadvantages of ability grouping, more schools are trying different approaches. In the US Carol Burris, Jay Heubert and Hank Levin conducted a study of a de-tracking innovation in mathematics[11]. The researchers compared 6 annual cohorts of students in a middle school in the district of New York. The students attending the school in 1995, 1996 and 1997 were taught in tracked classes with only high track students being taught the advanced curriculum. But in 1998, 1999 and 2000 *all students* in grades 7-9 were taught advanced curriculum in mixed ability classes and all of the 9th graders were taught an accelerated algebra course. The researchers looked at the impact of these different middle school experiences upon the students' completion of high school courses and their achievement, using 4 achievement measures, including scores on the advanced placement calculus examinations. They found that the students from de-tracked classes took more advanced classes, pass rates were significantly higher and students passed exams a year earlier than the average in New York State. The scores of the students were also significantly higher on various achievement tests. The increased success from de-tracking applied to students across the achievement range – from the lowest to the highest achievers.

Many parents fear mixed ability grouping and cannot see the logic of grouping students with very different needs and limited teacher resources into one class. So why is it that mixed ability grouping, is repeatedly found to be associated with higher achievement? The five most important reasons are these:

1. Opportunity to Learn.

Researchers consistently find that the most important factor in school success is what they call 'opportunity to learn'.[12] If students are not given opportunities to learn challenging and high level work then they do not achieve at high levels. We know that when students are in lower groups they receive low-level work and this, in itself, is damaging. In addition teachers inevitably have lower expectations for students. In the 1960's Robert Rosenthal and Leonore Jacobson, two sociologists, conducted an experiment to look at the impact of teacher expectations. Students in a San Francisco primary school were assigned to two groups. Both groups were taught the same work, but teachers were told that one group included all of the students identified as especially talented. In reality the students had been randomly assigned to the two groups. After the experimental period, the students in the group identified as talented gained better results and scored at higher levels on IQ tests. The authors concluded that this effect was entirely due to the different expectations the teachers held for the students.[13] In a study of 6 schools in England, I and a team of researchers were saddened to find that teachers routinely underestimated children in low groups, as the students reported to us:

> Sir treats us like we're babies, puts us down, makes us copy stuff off the board, puts up all the answers like we don't know anything.
>
> And we're not going to learn from that, 'cause we've got to think for ourselves.

The students' talked openly to us about the low expectations teachers held for them and the way their achievement was being held back, and observations of the classes confirmed this to be true. The students simply wanted to be given opportunities to learn:

Obviously we're not the cleverest, we're group 5, but still— it's still maths, we're still in year 9, we've still got to learn.[14]

When teachers have lower expectations for students and they teach them low-level work, the children's achievement is suppressed. This is the reason that ability grouping is illegal in many countries in the world, including Finland, the country that topped the world in the latest international achievement tests.[15]

2. High-Level Discussions

Students who struggle in mathematics, are helped by engaging in discussions about maths with students who are working at higher levels. Different research studies have found that when students work in mixed groups, the achievement of the high achieving students is unaltered but the achievement of the lower achieving students increases[16]. For the lower achievers, the bar is raised to the standard of the higher achievers. In situations where students who struggle are placed in groups with other students who struggle, they do not come into contact with high-level discussions and the bar is kept at a low standard for everyone.

3. Student Differences

When a student is placed into a setted group, high or low, assumptions are made about their potential achievement. Teachers tend to "pitch" their teaching to students in the middle of the group, and they teach a *particular* level of content assuming that all students are more or less the same. In such a system the work is, inevitably, at the wrong pace or the wrong level for many students within a group. The lower students struggle to catch up while others are held back. While a teacher of any class, including a mixed ability class, can (wrongly) assume that her students all have the same needs, setting *is based upon* this erroneous assumption and when teachers have a setted group they often feel at liberty to treat all students the same. This is true even when students clearly have different needs and would benefit from working at different paces. In conversations with teachers from Railside, I asked them about the way they dealt with groups of students who worked at different achievement levels in the same classes. One day I challenged the head of department at the time, saying – "Surely the students won't all achieve at the same levels on the tasks you give them?" No, he said, "they won't, but I refuse to predict in advance how well a student will do on any task – students continue to surprise and amaze me". The Railside teachers' unwillingness to prejudge students' work was a large part of the reason for the success of their students.

In a mixed ability group the teacher has to *open* the work, making it suitable for students working at different levels and different speeds. Instead of prejudging the achievement of students and delivering work at a *particular* level, the teacher has to provide work that is multi-leveled and that enables students to work at the highest levels they can reach. This means that work can be at the right level and pace for all students.

4. Borderline Casualties.

When teachers assign students to different groups they make decisions that affect their long-term achievement and their life chances. Despite the importance of such decisions, they are often made on the basis of insufficient evidence. In many cases students are assigned to groups on the basis of a single test score with some students missing the high groups because of one point. That one missed point, which students may have scored on another day, ends up limiting their achievement for the rest of their lives.

Researchers in Israel and the UK found that students on the borders of different groups had essentially the same understandings, but the ones entering higher groups ended up scoring at significantly higher levels at the end of school because of the group they were placed into. Indeed the group that students were put into was more important for their eventual achievement than the school they attended.[17] Borderline casualties are those students who just miss the high groups and become casualties of the grouping system from that day on.

In England we do not only have borderline casualties, we have large numbers of students who are *simply placed into the wrong groups*. Incorrect placement of students – whereby a school decides that a child's potential is lower than it really is, will always happen with ability grouping, but researchers conducting a national survey of 404 schools for the Department for Children, Schools and Families (DCSF) recently reported that *over half* of the students were incorrectly placed into groups[18]. This means that decisions that will determine the level of a child's success, not only in school, but for many years to come[19], were wrong in half of the cases, which is very worrying indeed.

5. Student Resources.

In a settled class the main sources of help are the teacher or the textbook. Students are presumed to have the same needs and to work in the same way so the teacher feels comfortable lecturing to the class for longer periods of time and requiring the class to work in silence or in very quiet conditions, denying them the many advantages of talking through problems, as set out in chapter 2.

In mixed ability classes the students are organized to work with each other and help each other. Instead of one person serving as the resource to 30 or more students, there are many. The students who do not understand as readily have access to many helpers. The students who do understand serve as helpers to classmates. This may seem like it is wasting the time of high achievers but the reason these students end up achieving at higher levels in such classrooms is because the act of explaining work to others deepens understanding. As students explain to others they uncover their own areas of weakness and are able to remedy them and they strengthen what they know. In the two longitudinal studies I conducted the high achieving students told me that they learned more and more deeply from having to explain work to others. These are some of the high achievers from Railside school, talking about their experiences in mixed ability groups in the US:

> Everybody in there is at a different level. But what makes the class good is that everybody's at different levels so everybody's constantly teaching each other and helping each other out. (Zane, Y2)

Some of the higher achievers started the school thinking it unfair that they had to explain work to others, but they changed their minds within the first year as they realized that the act of explaining was helping them, as these students reflect:

So maybe in ninth grade it's like Oh my God I don't feel like helping them, I just wanna get my work done, why do we have to take a group test? But once you get to AP Calc you're like Ooh I need a group test before I take a test. So like the more math you take and the more you learn you grow to appreciate, like Oh Thank God I'm in a group! (Imelda, Railside, Y4)

The high achievers also learned that different students could add more than they thought to discussions:

It's good working in groups because everybody else in the group can learn with you, so if someone doesn't understand – like if I don't understand but the other person does understand they can explain it to me, or vice versa, and I think it's cool. (Ana, Railside, Y3)

Students who work together, supporting each other's learning, provide a tremendous resource for each other, maximizing learning opportunities at the same time as learning important principles of communication and support.

'*I don't want to be a stupid person*'

When we consider the role of ability grouping, and the difference it makes in students' lives, there is another dimension, besides achievement, that it is critical to consider. For ability grouping not only limits opportunities, it influences the sorts of people our children will become. As students spend thousands of hours in their mathematics classrooms they do not only learn about mathematics, they learn about ways of acting and ways of being.

Mathematics classrooms influence, to a high and regrettable degree, the confidence students have in their own intelligence.

This is unfortunate both because maths classrooms often treat children harshly, but also because we know that there are many forms of intelligence and ways to be "smart" and maths classrooms tend to value only one. In addition to the power that maths classrooms have to build or crush children's confidence they also influence to a large extent the ideas students develop about other people.

Through my own research I have found that students in setted classes not only developed ideas about their own potential, but they began to categorize others in unfortunate ways – as smart or dumb, quick or slow. Comments such as these came from students in tracked (setted) classes in the US:

> I don't want to feel like a retard. Like if someone asks me the most basic question and I can't do it. I don't want to feel dumb. And I can't stand stupid people either. Because that's one of the things that annoy me. Like stupid people. And I don't want to be a stupid person.

The students who had worked in mixed attainment classes, at Railside school, did not talk in these ways and they developed impressive levels of respect for each other. Any observer to the classrooms could not fail to notice the respectful ways students interacted with each other, seeming not to notice the usual dividing lines of social class, ethnicity, gender or 'ability'. The ethnic cliques that often form in multicultural schools did not form at Railside and students talked about the ways their maths classes had taught them to be respectful of different people and ideas. In learning to consider different approaches to maths problems, students also learned to respect different ways of thinking more generally and the people making such contributions. Many of the students talked about the ways they had opened their minds and their ideas, for example:

T: You got everyone's perspective on it, 'cause like when you're debating it, a rule or a method you get someone else's perspective of what they think instead of just going off your own thoughts. That's why it was good with like a lot of people.

C: I liked it too. Most people opened up their ideas. (Tanita & Carol, Railside, Y4).

Undeniably one of the goals of schools is to teach students subject knowledge and understanding, but schools also have a responsibility to teach students to be good citizens – to be people who are open minded, thoughtful, and respectful of others who are different from themselves. The Carnegie Corporation of New York, based upon a report from the Council for Adolescent Development, recommended that de-tracking occur in schools in order to create 'environments that are caring, healthy and democratic as well as academic'.[20] The Coalition for Essential Schools also recommended that schools abandon tracking for both academic and moral reasons. Its director Ted Sizer argued that 'If we want citizens who take an active and thoughtful part in our democracy they must get trained for this in school – working together on equally challenging problems, and using every possible talent toward their solutions.'[21]

Although it seems to make sense to place students into groups where they have similar needs, the negative consequences of setting decisions, for students' achievement, and for their moral development, are too strong to ignore. Teachers of mathematics are often inexperienced with different systems of grouping but this is not a good reason for perpetuating a flawed approach. Teachers need to be supported in gearing their teaching towards high achievement for all. In the US I visited maths classes that had recently become de-tracked but teachers continued to use the same approach they had always used, to disastrous effect. These teachers gave out worksheets targeting small areas of content appropriate for a tiny number of the class, leaving the rest floundering, or bored.

For mixed ability classes to work well, two critical conditions need to be met. First, the students must be given open work that can be accessed at different levels and taken to different levels. Teachers have to provide problems that people will find challenging in different ways, not small problems targeting a small, specific piece of content. These are also the most interesting problems in mathematics and so they carry the additional advantage of being more engaging. I have seen such problems used to great effect in a number of classrooms. These sorts of multi-level problems are used in Japanese classrooms in order to promote high achievement for all, as Steve Olson, author of the best-selling book *Countdown* reflects:

'teachers *want* their students to struggle with problems because they believe that's how students come to really understand mathematical concepts. Schools do not group students into different ability levels, because the differences between students are seen as a resource that can broaden the discussion of how to solve a problem. Not all students will learn the same thing from a lesson; those who are interested in and talented at math will achieve a different level of proficiency from their classmates. But each student will learn more by having to struggle with the problem than by being force-fed a simple, predigested procedure.'[22]

Japanese students are not all expected to learn the same from each lesson, which is the unrealistic expectation in many countries; instead they are given challenging problems and each student gets the most from them that they can.

In addition to open, multi-level problems, the second critical condition for mixed ability classes to work, is that students are taught to work respectfully with each other. I have observed many maths classrooms where students are working in groups, but the students do not listen to each other. The teachers of such classrooms have given students good problems to work on together, and they have asked students to discuss the problems, but the students have not learned to work well together. This can result in chaotic classrooms with groups where only some of the students do the work, or, even worse, groups in which some students are ignored or ridiculed by other students because they are deemed to be "low status". Teaching students to work respectfully requires careful and consistent building of good group behaviour. Some teachers do this by highlighting the need for respect and hard work in groups, some teachers employ additional strategies such as "complex instruction", an approach designed for use with mixed attainment groups which is aimed at reducing status differences between students. Whatever the approach, when students learn to work well and respectfully together and their different strengths are seen as a resource rather than a point to ridicule, then children are helped by being able to achieve at high levels and society is helped through the development of respectful, caring young people.

Psychological Prisons

In my study of Amber Hill and Phoenix Park school in England I was able to follow students through a school that used ability grouping (Amber Hill) and one that did not (Phoenix Park). At Amber Hill the teachers also taught traditionally, whereas at Phoenix Park the teachers used complex, open problems. As I described in chapter 3 I was fortunate in being able to catch up with the students 8 years after my initial study, and talk to them about the impact of their school experiences on their jobs and lives. At that time the young people were around 24 years of age. During the follow-up study I found that one of the most important differences between the ex-students, perhaps not surprisingly, was the grouping approaches they had experienced. At Amber Hill, where they used ability grouping, the adults talked about how the grouping had shaped their whole school experience and many of those from set 2 downwards talked not only about the ways their achievement had been constrained by the grouping but also the ways they had been set up for low achievement in life. In a statistical comparison of the jobs that the ex-students were working in, I found that those who had experienced mixed ability grouping, despite growing up in one of the poorest areas in the country, were now in more professional jobs than those who had experienced ability grouping.[23]

Interviews with some of the young adults gave meaning to these interesting differences. The students from Phoenix Park talked about the ways the school had excelled at finding and promoting the potential of different students and that teachers had regarded everyone as a high achiever. The young adults communicated a positive approach to work and life, describing the ways they used the problem solving approaches they had been taught in school to get on in life. The young adults who had attended Amber Hill, that had put them into 'sets', told me that their ambitions were 'broken' at school and their expectations lowered. One of the young men spoke passionately about the ability grouping experience:

"You're putting this psychological prison around them (. . .), it's kind of . . . people don't know what they can do, or where the boundaries are, unless they're told at that kind of age."

"It kind of just breaks all their ambition (. . .). It's quite sad that there's kids there that could potentially be very, very smart and benefit us in so many ways, but it's just kind of broken down from a young age. So that's why I dislike the set system so much—because I think it almost formally labels kids as stupid." (Nikos, ex-Amber Hill student).   .

The impact that ability grouping has upon students' lives – in and beyond school – is profound. Researchers in England[24] found that 88% of children placed into ability groups at age 4 remain in the same groupings until they leave school. This is one of the most chilling statistics I have ever read. Children develop at different rates and stages, revealing different strengths and weaknesses at various stages of their development. For schools to decide what children can do, for the rest of their school career, when they are 4 years old – or any other primary school age – is nothing short of criminal. The single most important goal of schools is to provide stimulating environments for all children; with classrooms that offer opportunities and challenges for students, with teachers who are ready to encourage and nurture children's development – whenever it occurs. This can only be done through a flexible system of grouping that does not pre-judge a child's achievement and that uses multi-leveled mathematics materials that each student takes to their own highest level. Only such an approach will enable England to move away from notions of "low ability" children, to improve children's self-esteem, which is desperately needed, and to encourage smart and competent maths learners in our classrooms and our homes.

6 / Paying the price for sugar and spice?

How girls and women are kept out of maths and science.

When I started my research at Amber Hill Caroline was 14 years old, eager to learn and very accomplished. When the students had started the school some 3 years earlier, they had all been given a maths test. Caroline had earned the highest score in her year but 3 years after coming to the school she was the lowest achieving student in her class. How could it all have gone so badly wrong? When I met Caroline she had just been placed into the top set – a group that was taught by Tim, the friendly and well-qualified maths head of department. But Tim was a traditional teacher and he, like most maths teachers, demonstrated methods on the board and then expected students to work through exercises practicing the techniques. Caroline sat at a table of girls, six of them in total. All of the girls were high achievers and they all wanted to do well in maths. From the beginning Caroline looked uncomfortable in class. She was an inquisitive and thoughtful girl, and whenever Tim explained methods to the class, she, like many girls I have taught and observed over the years, had questions: Why does that method work? Where does it come from? And how does that fit with the methods we learned yesterday? Caroline asked Tim these questions, from time to time, but he would generally just re-explain the method, not really appreciating why she was asking. Caroline became less and less happy with maths and her achievement started to decline.

In one of their lessons the students were learning about the multiplication of binomial distributions. Tim had taught students to multiply binomial expressions such as:

$$(x + 3)(x + 7)$$

by telling students that they:

1. multiply the first terms (x times x)
2. multiply the outer terms (x times 7 and x times 3)
3. multiply the inner terms (3 times 7) and then
4. add all the terms together ($x^2 + 7x + 3x + 21 = x^2 + 10x + 21$)

Students are often taught to remember this sequence with the pneumonic FOIL (first, outer, inner, last). These are the sorts of procedures in maths class that often seem meaningless to students – they are hard to remember and easy to muddle up. I approached Caroline's group that day as she was sitting with her head in her hands. I asked her if she was alright, and she looked at me with an agonized expression. "Ugh I hate this stuff", she said, "can you tell me – why does it work like this? Why does it have to be in that order, with all of that adding?" She had tried asking Tim who had told her that that is how the formula works and you just need to remember it.

I knelt beside Caroline's desk and asked her whether I could draw a diagram for her. I explained that we could think about the multiplication visually by thinking of the two expressions as sides of a rectangle. She sat up and watched as I drew a sketch:

Soon all the girls at the table were watching. Before I had finished the drawing Caroline said "oooh, I see it now" and the others made similar appreciative noises. I felt a bit bad about offering this drawing, as my role in the classroom was not to help students and certainly not to undermine Tim's teaching, but it was an opportunity that I took on that one occasion. The drawing was simple but it offered a lot – it allowed the girls to see *why* the method worked and that was important for them.

I observed Tim's class, and other classes at the school, many times over the three years. As I interviewed more and more of the boys and girls I started to notice that the desire to know *why* was something that separated the girls from the boys. The girls were able to accept the methods that were shown to them and practice them, but they wanted to know *why* they worked, *where* they came from and how they *connected* with other methods. Some of the boys were also curious about the ways methods were connected and how they worked, but they seemed willing to adapt to a teaching approach that did not offer them such insights. In interviews the girls would frequently say such things as:

> He'll write it on the board and you end up thinking, well how comes this and this? How did you get that answer? Why did you do that? (Jane).

Many of the boys, on the other hand, would tell me that they were happy as long as they were getting answers correct. The boys seemed to enjoy completing work at a fast pace and competing with other students, and they did not seem to need the same depth of understanding. John spoke for many of the boys when he said:

> I dunno, the only maths lessons you like are when you've really done a lot of work and you're proud of yourself because you've done so much work, you're so much ahead of everyone else.

In a questionnaire I gave to the whole of the year group, I asked the students to rank five different ways of working in mathematics. Ninety-one per cent of girls chose 'understanding' as the most important aspect of learning mathematics, compared with only 65% of the boys, which was a statistically significant difference[1]; the rest of the boys said that the memorization of rules was the most important. The girls and boys also acted differently in lessons. During the hundred or so lessons that I observed I would often see boys racing through their textbook questions, trying to work as quickly as possible and complete as many questions as they could. I would, just as frequently, observe girls looking lost and confused, struggling to understand their work or giving up all together. In lessons I would often ask students to explain to me what they were doing. The vast majority of the time the students would tell me the chapter title and, if I asked them questions like 'yes but what are you actually *doing?*' they would tell me the number in the exercise; neither girls nor boys would be able to tell me why they were using methods or what they meant. On the whole the boys were unconcerned by this, as long as they were getting their questions right, as Neil told me in interview:

Some of the stuff you do, it's just hard and some of it's really easy and you can remember it every time; I mean, sometimes you try and race past the hard bits and get it mostly wrong, to go onto the easy bits that you like.

The girls would get questions right, but they wanted more, as Gill explained:

It's like, you have to work it out and you get the right answers but you don't know what you did, you don't know how you got them, you know?

At the end of the three years the students took their GCSE exams. In the top set that I had observed so closely, the girls achieved at significantly lower levels than the boys, which was a pattern that was repeated in different groups across the year cohort. Caroline, once the glittering star of the group, achieved the lowest grade of all. By the time she was finishing GCSE maths she had decided that she was no good at maths, despite the fact that she had shown a great aptitude for it some years earlier. Caroline, and many of the other girls, had underachieved because they had not been given the opportunity to ask why methods worked, where they came from and how they were connected. Their requests were not at all unreasonable – they wanted to locate the methods they were being shown within a broader sphere of understanding. Neither the boys nor the girls liked the traditional mathematics lessons at Amber Hill particularly – maths was not a popular subject at the school – but the boys worked within the procedural approach they were given whereas many of the girls resisted it. When they could not get access to the depth of understanding they wanted, the girls started to turn away from the subject. National statistics tell us that girls now do very well in mathematics, achieving at equal or higher levels than boys. This high achievement, given the inequitable approach that most girls experience, is testament to their capability and impressive motivation to do well. But the high achievement of girls often masks a worrying reality – the approaches they experience make many girls uncomfortable and their lack of opportunity to inquire deeply is the reason that so few girls continue with maths beyond GCSE.

At Phoenix Park school where students were taught through longer, more open problems and they were encouraged to ask why, when and how, the girls and boys achieved equally, and both groups achieved at higher levels than the students at Amber Hill, on a range of assessments, including national exams.

Some years later I was sitting in a maths class myself, having decided to take some higher level classes. We had a wonderful teacher, a woman who explained why and how methods worked – almost all of the time. I remember one day sitting in class when our teacher showed the formula for standard deviation and then said, by the way would you like to know why it works? Then a funny thing happened. The women in the class chorused "yes" and most of the men chorused "no". The women joked with the men asking: "What is wrong with you?!" One of the men responded quickly – "Why do we need to know why? It is better just to learn it and move on". It was then that I realized we were playing out the same gender roles as the students I had been observing at Amber Hill.

In one of my first research studies at Stanford University, I decided to learn more about the experiences of high achieving students in American high school classes. I chose six schools and interviewed 48 boys and girls about their experiences in calculus classes. In 4 of the schools the teachers used traditional approaches, giving the students formulas to memorize without discussing why or how they worked. In the other two schools the teachers used the same textbooks but they would always encourage discussions about the methods that students were using. I was not investigating or looking for gender differences but I was struck again by the reflections of the girls in the traditional classes and their need to inquire deeply, as Kate described:

K: We knew how to do it. But we didn't know why we were doing it and we didn't know how we got around to doing it. Especially with limits, we knew what the answer was, but we didn't know why or how we went around doing it. We just plugged into it. And I think that's what I really struggled with – I can get the answer, I just don't understand why. (Kate, Lime school)

Again many of the girls told me that they needed to know *why* and *how* methods worked and they talked about their dislike of classes in which they were just asked to memorize formulas, as Kristina and Betsy described:

K: I'm just not interested in, just, you give me a formula, I'm supposed to memorize the answer, apply it and that's it.
JB: Does maths have to be like that?
B: I've just kind of learned it that way. I don't know if there's any other way.
K: At the point I am right now, that's all I know. (Kristina & Betsy, Apple school)

Kristina went on to tell me that her need to explore and to understand phenomena was due to being a young woman:

> Maths is more like concrete, it's so "it's that and that's it." Women are more, they want to explore stuff and that's life kind of, like, and I think that's why I like English and science, I'm more interested in like phenomena and nature and animals and I'm just not interested in just you give me a formula, I'm supposed to memorize the answer, apply it and that's it. (Kristina, Apple school)

It was unfortunate for Kristina that mathematics was not one of her school subjects that allowed her to "explore" or to consider phenomena, when it should have been.

David Sela, from the ministry of education in Israel, and Anat Zohar from the Hebrew University in Jerusalem, conducted an extensive investigation of gender differences in the learning of physics[2]. They took my notion of the 'quest for understanding' that I had found to be prevalent among girls in maths classes and considered whether it was also prevalent among girls in physics classes. They found, resoundingly, that it was. The researchers drew from a database of approximately 400 high schools in Israel that offered advanced placement physics classes. They sampled 50 students from the schools and interviewed 25 girls and 25 boys. They found that the girls in physics classes were exhibiting the same preferences that I had found in mathematics classes, resisting the requirement to memorize without understanding, saying that it was "driving them nuts". The girls talked about wanting to know why methods worked and how they were linked. The authors concluded that 'although both girls and boys in the advanced placement physics classes share a quest for understanding, girls strive for it much more urgently than boys, and seem to suffer academically more than boys do in a classroom culture that does not value it.'[3]. Neither the female maths students I interviewed nor the female physics students interviewed in Israel wanted an easier science or maths, they did not need or want softer versions of the subjects, in fact the versions of mathematics and science they wanted required considerable depth of thought. In both cases the girls wanted opportunities to inquire deeply, and they were averse to versions of the subjects that emphasized rote learning. This was true of boys and girls but when girls were denied access to a deep, connected understanding, they turned away from the subject.

The differences that have been found between girls and boys in mathematics and physics classes do not suggest that all girls behave in one way and all boys in another. Indeed Zohar and Sela found that one third of the boys they interviewed also expressed strong preferences for a deep, connected understanding. But they, like I, found that girls consistently expressed such preferences in higher numbers and with more intensity. Such gender differences are interesting and, until recently, have been somewhat puzzling.

The idea that girls and women value a different type of knowing was famously proposed by Carol Gilligan, an internationally acclaimed psychologist and author. In Gilligan's book *In a Different Voice* she claimed that women are likely to be 'connected thinkers', preferring to use intuition, creativity, and personal experience when making moral judgments. Men, she proposed, are more likely to be 'separate' thinkers, preferring to use logic, rigour, absolute truth and rationality when making moral decisions.[4] Gilligan's work met a lot of resistance but it also received support from women who identified with the thinking styles she described. Some years later a group of researchers developed Gilligan's distinctions further, claiming that men and women differ in their ways of knowing, more generally. Psychologists Mary Belenky, Blythe Clinchy, Nancy Goldberger & Jill Tarule[5], proposed stages of knowing, and again claimed that men tended to be separate thinkers and women connected thinkers. The authors did not have a lot of data to support their claims that women and men think differently and they received considerable opposition, which is understandable given that they were suggesting fundamental distinctions in the way women and men come to *think* and *know*. When I reported my own findings, that girls were particularly disadvantaged by traditional instruction that did not give them access to knowing how and why, I also received resistance. Indeed some of my colleagues challenged me, saying that it was not possible that girls would have different preferences from boys in such a *cognitive* domain. They asked me for the reason for such differences and I admitted that it was not clear. Despite vastly different socialization processes that girls and boys are subjected to from an early age, there seemed to be no obvious reason for their different preferences in maths and science knowing. That is, until now.

Emerging Research on the Brain.

One of the most interesting developments of the 21st century has been a new form of technology that is allowing neuroscientists to map the workings of the human brain. Louann Brizendine, a neuropsychiatrist and author of the book *The Female Brain*, explains that brain-imaging technology has allowed scientists to document "an astonishing array of structural, chemical, genetic, hormonal and functional brain differences between men and women."[6] Researchers have now found that women and men use different brain areas to solve problems, even when they score exactly the same on tests. For example, researchers found that when participants were asked to mentally rotate three-dimensional shapes they were equally good at it but women and men used completely different brain circuits.[7] Some of the most interesting findings from brain-imaging research are these:

1. While scientists have known for hundreds of years that men's brains are bigger than women's[8], even when controlling for men's greater average size, they now know that women and men have exactly the same number of brain cells and so have equal intelligence. In men's brains the cells are packed less densely.

2. A huge testosterone surge beginning in the eighth week of pregnancy kills cells in the communication centers of male brains and grows more cells in the sex and aggression centers. In the brain centers for language and hearing, for example, women have 11% more neurons. The fact that autism, a communication disorder, affects eight times as many boys as girls is thought to be due to the fact that men's brains offer less support for communication[9].

3. From the first days of birth girl babies are more attuned to faces and human interactions. In a study conducted by graduate student Jennifer Connellan and her Cambridge University professor Simon Baron-Cohen, that compared newborn babies *on the day they were born*, babies were given a choice between looking at a simple dangling mobile or a young woman's face. One hundred and two babies in the study were videotaped and their eye motions were analyzed by researchers who did not know the sex of the babies. Boy babies were almost twice as likely to look at the mobile than girl babies, who preferred to look at the women's face. The researchers concluded "beyond reasonable doubt" that sex differences in social interest must, in part, be biological.[10]

4. Brain function in men is more compartmentalized. In men the left side of the brain is specialized for language function but women's brains are more symmetrical.

Research on stroke victims shows that when men have a stroke involving the brain's left hemisphere they drop their verbal IQ by about 20 percent whereas those who have a stroke involving the brain's right hemisphere do not drop any verbal IQ points. Women who have a stroke involving the brain's left hemisphere drop their verbal IQ by about 9 percent, those who have a stroke involving the brain's right hemisphere drop their verbal IQ by about 11 percent. Scientists conclude from this that women use both sides of their brain for language and men do not.

5. In a study at Stanford University, Brizendine reports, researchers studied the brains of volunteers who were viewing emotional images. The brain scans showed that nine different brain areas lit up in women, but only two in men.[11]

6. Brain imaging studies have found that for many tasks, including mathematical work women use the more advanced area of the brain – the cerebral cortex, whereas men use the more "primitive" areas of the brain such as the globus pallidus, the amygdala, or the hippocampus."[12]

These and other striking findings lead Brizendine to conclude that the girl brain is "a machine that is built for connection". Leonard Sax, a physician and author of the book *Why gender matters: what parents and teachers need to know about the emerging science of sex difference* also reviews the research on brain differences and draws similar conclusions to Brizendine. Sax argues that girls and boys hear differently (and he presents evidence that girls hear better than boys), play differently and *learn differently*. Part of the evidence that Sax gives for his claim that girls and boys learn differently is that brain imaging studies have found that men and women use different areas of the brain when they complete tasks. Sax concludes that the male and female brains are equally powerful but organized differently and suited to different tasks.

Both of the authors conclude that women and girls are naturally inclined towards communication and connection making and they both draw conclusions about the ways girls and boys should learn maths and science, but only one author offers conclusions that are supported by educational research. Louann Brizendine makes a point that is supported by findings on successful and less successful teaching environments. She says, quite rightly, that girls often reject mathematics because they want to be learning subjects that involve more communication and more connections with people. At a time when young people are choosing career paths, the male brain, Brizendine explains, is flooded with testosterone and boys become content with more solitary pursuits. At the same time young women often look to work with people and to learn collaboratively. These female choices become incompatible with the way mathematics is generally taught, *but not the way it should be taught*. When I interviewed students in calculus classes in the US, the girls who were taught in classes that encouraged the discussion of concepts saw mathematics as compatible with their preferences for communication, understanding and depth, as Veena, who planned to take a degree in mathematics, commented:

> Sometimes you sit there and go 'it's fun!' I'm a very verbal person and I'll just ask a question and even if I sound like a total idiot and it's a stupid question I'm just not seeing it, but usually for me it clicks pretty easily and then I can go on and work on it. (. . .) There's this certain point when it just connects and you see the connection and you get it.

Classes in which students discuss concepts, giving them access to a deep and connected understanding of maths are good for girls and for boys. Boys may be willing to work in isolation on abstract rules but such approaches do not give many students, girls or boys, access to the understanding they need. In addition, high level work in mathematics, science and engineering is not about isolated, abstract rule following, but about collaboration and connection making.

Sax draws different conclusions about the implications of differences in brain functioning for mathematics learning, that are not supported by educational research. He notes that brain imaging has shown that when girls work on maths problems they are using the cerebral cortex, which is the part of the brain that mediates language and higher cognitive functions, whereas boys use the hippocampus, 'a phylogenetically primitive area of the brain that is pre-wired for spatial navigation.'[13] Sax points out, quite correctly, that girls need maths to be tied into higher cognitive functions, but his suggestion is that boys are taught using raw numbers, isolated from contexts, whereas girls are taught the same maths using real world contexts. He gives the teaching of Fibonnaci numbers as an example with the boys being given the fascinating sequence to explore and girls being asked to bring in objects from the world such as artichokes, sunflowers and pinecones. Sax proposes that girls would then ask questions such as: 'why do numbers in the Fibonacci sequence keep showing up when you count the petals on a delphinium' and 'How can abstract number theory explain these similarities?' He is right about the *why* and *how* questions girls may ask but wrong in suggesting that girls need real world examples. The reality is that girls would enjoy exploring sequences of numbers just as much as the boys, if they were inquiring deeply about them, and boys would enjoy considering the real world phenomenon. A good approach to the teaching of mathematics would involve *both aspects of mathematical work – abstract and applied*. Sax suggests putting boys into one class and giving them a disconnected (and distorted) approach and girls into another to get a connected but equally distorted version of mathematics. Even if girls did perform better when concepts were introduced via real world phenomenon, the mathematics curriculum would be pretty sparse if it only involved concepts that could be illustrated in such ways. The fact is that girls enjoy playing with numbers and number sequences and they do not need them to be embedded in the real world, if the concepts are connected to *other mathematical concepts*, and if there is discussion of *how* and *why* they work.

In addition, brain differences are not so compelling that they justify teaching girls and boys separately. We know that brain differences only explain gender tendencies and that there are plenty of boys who value and need connections and communication and who choose other subjects because mathematics does not offer these, just as there are girls who can happily work in isolation without mathematical connections. I would pity those girls and boys who did not conform to the ways of thinking and working that were typical for their gender but who were forced to be taught via a distorted version of mathematics that was entirely abstract or entirely based in the real world. If mathematics teaching included opportunities for discussion of concepts, for depth of understanding and for connecting between mathematical concepts, then it would be more equitable, good for both sexes, and it would give a more accurate depiction of mathematics as it is practiced in high level courses and professions.

Brain research is still in its infancy but the question of *why* women and girls want to inquire deeply is not as important as the question of how we provide environments in which they can.

Where are we now?

Given the ways in which mathematics is commonly taught and the preferences of many girls for deep understanding and inquiry, it is perhaps surprising that girls do as well as they do in mathematics – and they do perform very well. In 2006, 57% of girls and 55% of boys achieved A*-C grades at GCSE in England, and 83% of girls who took A-level mathematics achieved an A-C grade, compared with 78% of boys[14]. These statistics may be surprising as there is widespread public belief, fuelled by misleading media headlines, that boys are way ahead of girls in mathematics, which is far from the truth. Psychologists Janet Hyde, Elizabeth Fennema and Susan Lamon produced a meta-analysis[15] of studies that have investigated gender differences in achievement, combining over 100 studies involving 3 million subjects. Even in 1990, with such a vast database, they found very small differences between girls and boys, with a huge amount of overlap[16]. Hyde and her colleagues argued that gender differences were too small to be of any importance, and that they have been overplayed in the media which has helped to create stereotypes that are damaging.

Girls are doing very well now, but their strong performance hides a worrying fact – most mathematics classrooms are not equitable environments and girls often do well despite inequitable teaching. The inequities that remain in mathematics teaching in England are reflected in shockingly low rates of A-level participation among young women. In 2006 only 38% of A-level students were girls, despite their high performance at GCSE. In the United States, by contrast, 48% of math majors (earning a degree in mathematics) were women, even in 2002. The participation rates for young women and men in A-level mathematics in England show that we still have a long way to go before we can claim gender equity in mathematics. In Europe the low numbers of women working as researchers and scientists is one of the priority areas for the European Union. Fifty-two per cent of higher education graduates across Europe are women but only 25% of these women take science, engineering or technology subjects. Girls do well in maths and science because they are capable and conscientious but many do so through endurance, and maths classroom environments are far from being equitable. Indeed it is the impoverished version of mathematics that is offered to students that turns many people, female and male, away from the subject.

Other Barriers.

Of course the lack of opportunity to inquire deeply is not the only barrier to girls and women in mathematics and science. Mathematics classrooms in schools are considerably less gender stereotyped than they were twenty years ago, when sexist images prevailed in textbooks and mathematics teachers were found to give boys more attention, reinforcement and positive feedback,[17] but still girls in some classrooms experience stereotyped attitudes and behaviours, contributing to their low interest and participation in maths. Some school mathematics and science classrooms are also highly competitive which deters many young women.

In university maths departments the situation is worse. Abby Herzig, a professor from the State University, New York, has produced evidence of the ways in which the climates of university mathematics departments can be icy cold towards women and ethnic minority students[18]. Herzig notes that women face many issues including sexism, stereotyped ideas about women's capabilities, feelings of isolation, and lack of role models[19], especially at the post-graduate level. By and large, university mathematics departments remain a male preserve where the under-representation of women among students is eclipsed only by the under-representation of women among the faculty. At Stanford University there used to be no women's toilets in the mathematics department. In 2007, the department had still not built women's toilets, but they had added WO to the 'men' sign on some doors and put pots of flowers in the urinals. What sort of message does that send to the women taking courses in mathematics? There could not be a clearer statement about the absence of women in the history of the department, or the lack of concern for their sense of inclusion now. The teaching in mathematics departments can also be highly rule-based which again denies girls the opportunity to ask why and how, as Julie, one of the young women who had given up on her mathematics degree at Cambridge university, explained to me:

> I think it was my fault because I did want to understand every single step and I kind of wouldn't think about the final step if I hadn't understood an in-between step (. . .) I couldn't really see why they, how they got to it. Sometimes you want to know, I actually wanted to know.

For Julie, who had won awards for her mathematics achievement prior to university, her desire to understand how and why methods worked, stopped her from going forward with the subject.

Societal stereotypes.

In addition to the problems in university mathematics departments and schools, young women and men suffer from the stereotypes that are perpetuated in society, particularly by the media. Ideas such as girls being too nurturing and caring to work in the "hard sciences" are based upon incorrect ideas about women and about the ways mathematics and sciences work. Girls and women do not need a softer version of mathematics. Indeed the sorts of inquiries women need could be said to epitomize true mathematical work, with its need for proof and rigorous analysis of ideas.

When girls went ahead of boys in mathematics and science (and all other subjects) in England alarm bells rang everywhere. Suddenly there was government money to look into gender relations in maths and science, something that had never happened when girls were under achieving. It was interesting that whereas people had always decided that the under achievement of girls in maths and science was due to their intellect, when it was boys who were underachieving people looked to external reasons – with suggestions that books must be biased, the teaching approaches favoured girls, or that teachers must be encouraging girls more. Nobody suggested that boys did not have the brains for maths or science. Michele Cohen, a historian, gives an interesting perspective on the tendency of people to locate girls' underachievement *within* girls. She points out that this has been done throughout history and that in the 17th century scholars went to enormous lengths to explain away the achievement of girls and the working classes, as it was boys, specifically upper-class boys, that were believed to possess true intellect. People at that time explained that the superior verbal competence noted amongst women was a sign of weakness and that the English gentleman's reticent tongue and inarticulateness was evidence of the depth and strength of his mind. Conversely, women's conversational skills became evidence of the shallowness and weakness of their mind.[20] In 1897 the Reverend John Bennett, from the Church of England, argued that boys appeared slow and dull because they were thoughtful and deep and because 'gold sparkles less than tinsel.'[21]

The tendency to locate sources of underachievement within girls and to construct ideas about female inadequacy is also characteristic of much of the psychological research on gender. In my interviews with school students, I frequently encounter stereotypes about the potential of girls and boys. But it is particularly disturbing when I find that ideas about girls being mathematically inferior have come from the reporting of research. In a recent interview with a group of high school students in California, I asked Kristina and Betsy about gender differences:

JB: Do you think math is different for boys and girls or the same?

K: Well, it's proved that boys are better in math than girls, but in this class, I don't know.

JB: Mmm, where do you hear that boys are better than girls?

K: That's everywhere – that guys are better in math and girls are like better in English.

JB: Really?

B: Yeh I watched it on 20:20 [a television current affairs program] saying girls are no good, and I thought – well if we're not good at it, then why are you making me learn it?

The girls refer to a television programme that presented the results of research on the differences between the mathematical performance of girls and boys. The problem with such programmes, and some research analyses, is not that researchers noted that girls were achieving less than boys, or that girls displayed less confidence in maths classrooms, but that such findings were often presented as being due to the nature of girls, rather than any external sources. This led educators to propose interventions that were well intentioned but that aimed to change girls. The 1980's spawned numerous programmes that were intended to make girls more confident and challenging.[22] The idea behind such programmes is often good, but they also lay the responsibility for change at the feet of the girls rather than mathematics teaching environments or the broader social system. On July 5th 1989, the New York Times ran the headline: "Numbers Don't Lie: Men Do Better than Women" with the sub-heading "S.A.T. Scores accurately reflect male superiority in math." But this article, like many others, used performance differences to suggest that women were mathematically inferior, rather than questioning the teaching and learning environments as well as the biases they themselves were helping to create, that caused the under achievement of women. Now that women are ahead in most areas it is interesting to note the absence of any analogous headlines proclaiming women's innate superiority.

It was once suggested that girls are made of 'sugar and spice and all things nice' – a harmless fairy tale perhaps, but the idea that girls are sugary sweet and lack the intellectual rigor for mathematics and science is still around. It is time that such ideas are buried and that girls are encouraged to go into maths and science, for their sake as well as the sake of the disciplines themselves. Mathematics is and has always been about deep inquiry, connection making and rigorous thought. Girls are ideally suited to the study of high level mathematics and the only reason that they are under-represented now is because the subject is misrepresented and taught badly in too many classrooms.

7 / Talking with Numbers

Key Strategies and Ways of Working

All parents are faced with a challenging task – of finding the best ways to help their children develop, in social, emotional, moral, spiritual and intellectual ways. As a mother of two young children myself I know how daunting this challenge is. For many parents one of the most difficult barriers they face is their children's relationship with mathematics, especially if children have negative experiences in school, which many do. And this relationship is something to be taken very seriously as mathematics, as I have discussed in earlier chapters, has the power to crush children's confidence, with negative experiences in school not only making children feel inadequate and stupid but denying them access to a subject they will need for the rest of their lives. Fortunately there are some clear principles about learning mathematics that could help parents greatly, both in early work with very young children or work with older children who are having a hard time in school. In this chapter and the next I will share these principles and ways of working.

In this chapter, which will serve as essential background also to the next, I will review a research study that found that one particular way of working was *absolutely critical* to success, separating those who were achieving in mathematics from those who were failing. In the first half of the chapter I will explain this way of working carefully. In the second part I will finish with a description, from my own research and teaching, of the time that my post-graduate students and I set out to teach this and some other critical ways of working to some underachieving students who, like many others, had sat through years of maths classes without ever learning how to work in productive ways. I include this description here because I think the details may be helpful for parents. We had only 5 weeks with the students, in the challenging setting of summer school, and what we achieved with the students is eminently achievable by parents working with children over longer periods of time at home. I also include the stories of some children who changed as a result of the summer school, as their different circumstances and reasons for resisting mathematical work may remind readers of their own children and the barriers they may be facing in school. In the next chapter I will turn to the ways that parents may introduce the same strategies with their own children at home, as well as the ways they may work with teachers and schools to bring about improvements.

1. A Critical Way of Working.

In an influential research study, published in 1994[1], two researchers from the University of Warwick, Eddie Gray and David Tall, identified the reasons why many children struggle with maths. The results were so important that they should be shouted from the rooftops and posted in every maths classroom across the UK.

Gray and Tall conducted a study of 72 students between the ages of 7 and 13. They asked teachers in England to identify children from their classes who they regarded as 'above average', 'average' or 'below average' and they interviewed the 72 children. The researchers gave children various addition and subtraction problems to do. One type of problem required adding a single digit number, such as 4, to a teen digit number such as 13. They then recorded the different strategies children used. These strategies turned out to be critical in predicting children's achievement.

For example, let's take 4 + 13.

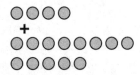

One strategy for solving this addition problem is called 'counting all'. With this strategy a student would look at the 4 dots and count them (1-2-3-4); they then look at 13 dots and count them (1-2-3-4-5-6-7-8-9-10-11-12-13), they then look at all the dots and count them, from 1 to 17. This is often the first strategy that children use as they are learning to count.

A more sophisticated strategy that develops from 'counting all' is called 'counting on'. A student using this strategy would count from 1 to 4. They would then continue on from 5 to 17.

A third strategy is called 'known facts' – some people just know, without adding or thinking, that 4 and 13 is 17 because they remember these number facts.

The fourth strategy is called 'derived facts'. This is where students decompose and recompose the numbers to make them more familiar numbers for adding and subtracting. So they may say – well I know 10 and 4 is 14 and then they add on the 3.

This sort of strategy of decomposing and recomposing numbers is really helpful when being given calculations to do, especially when doing them in your head. For example you may need to know the answer to 96 +17. To most people that is a nasty addition sum that looks daunting, but if 4 is taken from the 17 first and added to the 96, the problem becomes 100 +13 which is much more reasonable. People who are good at mathematics decompose and recompose numbers all the time. This is the strategy that the researchers called 'derived facts' as the students changed the numbers into ones that they knew the answers to, by decomposing and then recomposing the numbers.

The researchers found that the *above average* children in the 8+ age group 'counted on' in 9% of the cases, they used 'known facts' 30% of the time, and they used 'derived facts' 61% of the time. In the same age group the students who were *below average* 'counted all' 22% of the time, 'counted on' 72% of the time, used known facts 6% of the time, and they *never* used derived facts. It was this absence of derived facts that was critical to their low achievement.

When the researchers looked at 10 year olds they found that the 'below average' group used the same number of known facts as the above average 8 year olds, so you could think of them as having learned more facts over the years but, noticeably, they were still not using derived facts, instead they were still counting. What we learn from this, and from other research, is that the high achieving students don't just know more but they work in very different ways – and critically, they engage in flexible thinking when they work with numbers, decomposing and recomposing numbers.

The researchers drew two important conclusions from their findings. One was that low achievers are often thought of as *slow* learners, when in fact they are not learning the same things slowly, they are learning *a different mathematics*. The second is that the mathematics that low achievers are learning is a *more difficult* subject.

As an example of the very difficult mathematics that the below average children were using consider the strategy of 'counting back' which they frequently used with subtraction problems. For example, when they were given problems like 16 − 13 they would start at the number 16 and count down 13 numbers (16-15-14-13-12-11-10-9-8-7-6-5-4-3). The cognitive complexity of this task is enormous and the room for mistakes is huge. The above average children did not do this, they said 16 take away 10 is 6 and 6 take away 3 is 3, which is much easier. It seemed from the research that the students who were achieving at high levels were those who had worked out that numbers can be flexibly broken apart and put together again. The problem for the low achieving children was simply that they had not learned to do this. The researchers also found that when low achievers failed at their methods they did not change their method, instead they fell back into counting more and more. Indeed many of the low achievers became very efficient with small numbers, which lulled them into a sense of security. The low achieving students came to believe that in order to be successful they needed to count very precisely. Unfortunately problems become more and more difficult in mathematics and as time went on the low achievers were trying to count in more and more complex situations. Meanwhile the high achievers had forgotten counting strategies and were working with numbers flexibly. This is an easier task but it is also a more important way of working in mathematics − as the low achievers continued to count, the high achievers worked flexibly and pulled further and further ahead.

Not surprisingly, the researchers found that the lower achieving students who were not using numbers flexibly, were also missing out on other important mathematical activities. For example, one of the important things that people do as they learn mathematics is to compress ideas. What this means is, when we are learning a new area of maths, such as multiplication, we may initially struggle with the methods and the ideas and have to practice lots of examples, but at some point things become clearer, at which time we compress what we know and move on to harder ideas. At a later stage when we need to use multiplication, we can use it fairly automatically, without thinking about the process in depth.

William Thurston is a professor of mathematics and computer science at Cornell University who won the highest honour awarded in mathematics – the Fields Medal. He described the process of learning mathematics well:

> Mathematics is amazingly compressible: you may struggle a long time, step by step, to work through the same process or idea from several approaches. But once you really understand it and have the mental perspective to see it as a whole, there is often a tremendous mental compression. You can file it away, recall it quickly and completely when you need it, and use it as just one step in some other mental process. The insight that goes with this compression is one of the real joys of mathematics.[2]

One way of thinking about the learning of mathematics, visually, is to think of an upturned cone shaped light with the circular area of light being the area we are working in now, and the narrower sections being the mathematics we have learned before and we have compressed.

It is this compression that makes it easy for people to use concepts they learned many years ago, such as addition or multiplication, without having to think about how they work every time they use them. Gray and Tall found that the lower achieving students were *compressing* ideas less – they were so focused on remembering their different methods, and stacking one new method on top of the next, that they were not thinking about the bigger concepts and compressing the mathematics they were learning. For the low achieving students the learning of mathematics was less like an expanding light and more like a never-ending ladder, stretching up to the sky, with every rung of the ladder being another procedure to learn.

Students who are not taught to flexibly use numbers often cling to methods and procedures they are taught, believing that each method is equally important and must simply be remembered and reproduced carefully. For these students the mathematics they are learning is *much* more difficult.

Low achieving students who struggle to master more and more procedures, without using numbers flexibly, or compressing concepts, are working with the wrong model of mathematics. These students need to work with someone who will change their worldview of mathematics and show them how to use numbers flexibly and how to think about mathematical concepts. But instead of working with people who will change the students' approach, what typically happens is that students get labelled as low achieving students and people decide they need more practice, putting them into classes where they repeat methods over and over again. This is the last thing these students need and it simply feeds into their faulty world view of maths. Instead they need opportunities to play with numbers, as I will suggest in the next chapter, and to develop "number sense". Fortunately students can learn to use numbers flexibly and to consider concepts, *at any age*, and it was with this knowledge that my doctoral students and I set out to work with students who had previously been low achievers and to give them the experience of using mathematics flexibly. This is the story of what happened.

The Summer of Maths.

As I prepared to teach students in 7th and 8th grade in Californian classrooms I was a little nervous – it had been some years since I had taught mathematics in schools and all of my experience had been in London. The setting was a 5-week summer school class in the San Francisco Bay Area. Summer school classes are not known for their serious mathematics work and the students are there for a short time so it is difficult to establish careful and good classroom routines. The classes were also extremely mixed, combining students who loved maths and wanted to spend more time with the subject at the end of regular school, with those who had been forced to attend because they were failing in school. Reflecting this, 40% of the students had gained A's or B's in the previous year and 40% had gained D's or F's (these are known as failing students in the US).

Four of my doctoral students – Nick, Tesha, Emily, Jennifer – and I taught 4 classes of 7th and 8th graders for 2 hours each day, four days a week. Our teaching, and the students' learning, was the focus of a research study – lessons were observed and students were given surveys, interviews and assessments to monitor their learning. The classes were diverse both racially (39% Hispanic, 34% White, 11% African-American, 10% Asian, 5% Filipino, 1% Native American,) and socio-economically. I will start with a brief description of the summer school teaching and then illustrate the impact of the teaching by recalling the experiences of 4 very interesting young people.

One of the goals of our summer teaching was to give students opportunities to use mathematics flexibly and to learn to decompose and recompose numbers. We also wanted students to learn to ask mathematical questions, to explore patterns and relationships, and to think, generalize and problem solve, all of which are critical ways of working in mathematics, yet often neglected in school classrooms. As most of the students had spent the last year ploughing through worksheets rehearsing procedures in maths classes, this was quite a change for them. We decided to focus the summer school on algebraic thinking, as we thought that would be most helpful to the students in future years, and to focus upon critical ways of working including: asking questions, using mathematics flexibly, reasoning and representing ideas. These different ways of working are all critical to success in mathematics but are often overlooked in classrooms.

It would be reasonable to assume that students, particularly those in 7th and 8th grade, know how to ask questions in a maths class. Question asking is an important part of being a learner and one of the most useful things a student can do. But research has shown that student questions decline as students move through school and are surprisingly rare in classrooms.[3,4] Indeed such research suggests that as students progress through school, they learn *not to* ask questions and to keep silent, even when they don't understand. The act of question asking has been shown to increase mathematics achievement and improve attitudes among students, and we know that students who ask a lot of questions are usually the highest achievers.[5] But while many teachers regard student questions as valuable, they do not explicitly encourage questions. In our teaching of the summer school classes we chose to encourage student questions and to teach students the qualities of a good mathematics question. We started our sessions by telling students how much we valued questions and when the students asked good questions we would post them onto large pieces of paper that we hung up around the room. We also gave students mathematical problems and encouraged them to extend the problems by posing their own questions. When the students were interviewed during and after the summer many of them mentioned that they had learned that question asking was one of the most useful strategies in maths classes.

A second aspect of maths learning that we encouraged was that of mathematical *reasoning*. Students learn to reason through being asked, for example, to justify their mathematical claims, explain why something makes sense, or defend their answers and methods to mathematical sceptics.[6,7,8,9] Students who learn to reason about situations and determine whether they have been correctly answered learn that mathematics is a subject that they can make sense of, rather than being a list of procedures to memorize. When we talked to students, individually, in groups or to the whole class, we would always ask students to tell us why they thought an answer made sense and to justify it to their peers. By the end of the summer students were doing this for themselves, pushing each other to explain and justify when students talked about their mathematical ideas.

In addition to question asking and reasoning we also highlighted the importance of mathematical representations. Proficient problem solvers frequently use representations to solve problems and communicate results. For example, they may transform a problem given in numbers into a graph or diagram that illuminates different aspects of the problem. Or they may choose a particular representation to highlight something that helps a collaborator understand better. Professionals in mathematical jobs such as engineering or nursing frequently represent their ideas as part of their work,[10,11] both in understanding and communicating. Although representation is a critical part of mathematical work and it is often the first thing that proficient problem solvers do, it is rarely taught in classrooms. In our summer school teaching we gave the students problems that were displayed in different ways, particularly visually, with models and diagrams, and we asked the students to produce representations as part of their work. In interviews with the students some of them told us that they had *never* seen a mathematical method or idea represented visually before, and that the different representations they had seen and learned had been extremely powerful for them.

In all of our problems we encouraged and valued the flexible use of numbers. One of the best methods I know for teaching students how to use numbers flexibly is an approach called 'number talks', devised by leading educator, Ruth Parker. In number talks the teacher asks students to work individually and without paper and pen or pencil. The teacher puts a calculation (usually an addition or multiplication problem) onto the board or overhead and asks students to work out the answer in their heads. An example of a problem we posed was 18×5. Teachers ask the students to signal privately when they have an answer, usually by showing one thumb, but not by raising their hand as this is too public and puts other students under pressure, also turning the activity into a speed contest. The teacher then collects all the different methods students have used. When we posed 18×5 four students shared their different methods for working it out. They were:

$$18 \times 5$$

18 + 2 = 20	10 × 5 = 50	15 × 5 = 75	18 × 5
20 × 5 = 100	8 × 5 = 40	3 × 5 = 15	= 10 × 9
5 × 2 = 10	50 + 40 = 90	75 + 15 = 90	10 × 9 = 90
100 − 10 = 90			

These different methods all involve decomposing and recomposing numbers, changing the original calculation into other equivalent calculations that are easier. As students offered these examples, others saw a flexible use of numbers for the first time.

As the students took part in number talks they realized that there was no pressure to finish quickly and that they could use any method they were comfortable with, and they began to enjoy them a lot. Some of the students learned, for the first time ever, to decompose and recompose numbers, as Gray and Tall recommended, as they saw others doing this and realized that it was extremely helpful. Many of the students mentioned the number talks as a highlight of the summer, as they particularly liked the challenge, and the experience of sharing and seeing different mathematical methods. Although mathematics has a reputation for being a subject of single methods – with each problem requiring one standard method that must be remembered, nothing could be further from the truth. Part of the beauty of mathematical problems is that they can be seen and approached in different ways and although many have one answer, they can be answered using different approaches. When we asked students about aspects of the teaching that had been helpful, learning about different methods was one of the most frequently cited aspects, second only to collaboration with class-mates.

On the first day of summer school we gave students a blank journal to develop and record their mathematical thinking during the five-week course. We wanted the journals to be a space for students to play with ideas, as well as a safe place to communicate with us, for those who were afraid or less willing to share ideas publicly. We collected the journals frequently to look for mathematical thinking that we could help with and to give students feedback. Few of the students had ever been given the opportunity to write about maths before, nor to keep organized notes, which turned out to be very important for some of the students.

In most of the lessons students worked together in small groups or with partners. For some lessons we allowed students to choose their seats and groups; at other times we chose seating arrangements to help students work productively and to give them experience of working with a range of people and ideas. Our decision to allow students to sit where they wished at times was part of our general commitment to the promotion and encouragement of student choice, autonomy and responsibility. We emphasized to students that they were in charge of their own learning, encouraging them to make changes in their behaviour to improve their learning in class.

During the summer we combined different tasks and ways of working, as variety is very important in mathematical work. There are many valuable ways to work in maths classes – including through lectures, student discussions, and individual work, and there are many important types of tasks that students can work on, from long applied projects to short questions including contextual and abstract investigations. But none of these methods or tasks should be used exclusively as there is benefit to students experiencing *a range* of ways of working, especially as they will need to work in different ways in their jobs and lives. One of the biggest and surely the most reasonable complaints of students who experience traditional maths classes is that they are always the same. The monotony causes disaffection, it also means students only learn to work as they have in class – using procedures that have just been shown to them. During our classes we spent some time discussing ideas as a class and some time discussing ideas in groups. We often gave the students long problems to work through with others, and sometimes gave them shorter questions on worksheets to work through alone. We gave them tasks in which they explored patterns, similar to the algebraic work given at Phoenix Park and Railside, and we also gave tasks that were applied, such as a football world cup activity. In that task students needed to work out which teams would play each other and how many different games there would be, as an introduction to combinatorics. Sometimes students were given tasks to work on for a set amount of time, sometimes we allowed students to choose the tasks they worked on and to choose the amount of time they spent on them. Students were also encouraged to use their own ideas in extending problems and choosing methods that made the most sense to them. In all of our activities we encouraged students to ask questions, to represent, reason, and generalize and to share and think about different methods. We also spent a lot of time building students' confidence, praising them for their work and their thinking when it was good.

At the end of our classes we gave students the same algebra test that had been given them some months earlier in their regular classes. The students scored at significantly higher levels, even though we had not seen or taught to the content of the tests and our teaching had been significantly broader than the content of the tests. The average score before the summer was 48%, at the end of the summer it was 63%. In surveys 87% of students reported that the summer classes were more useful to them than their regular classes and 78% reported that they had enjoyed the classes 'a lot'. In interviews all of the students were extremely positive about the summer classes. Many of the students had told interviewers that their school classes were boring and frustrating, partly because they were made to work in silence. When reflecting on the summer class the students talked about their enjoyment and learning, in particular from the collaboration with peers, from learning about multiple methods, from the different strategies they learned, and the opportunities they received to think and reason.

In addition to the positive reports given in interviews and anonymous surveys, there was a huge improvement in the participation of the students in class during the summer. In the first class we asked students to complete a survey that asked them whose idea it had been to come to summer school and whether or not they wanted to be there. This revealed that 90% of students had not chosen to be there and most of them said that they did not want to be there. The main reasons given by those who said they did not want to attend were that it would be boring, that they were losing their summer, that they would rather be socializing with friends and that it would be unnecessary. The students' initial participation in our classes reflected their lack of enthusiasm. In the first session, many of the students were either quietly withdrawn, sitting with heads on their arms or 'hiding' under hoods, or they socialized with friends, chatting loudly and resisting our requests to work. We were thrilled that as the summer progressed, students' participation changed dramatically. After only a few days, students began to arrive at the door of class enthusiastic to start, they were taking the maths problems very seriously, they were interested in mathematical questions; they participated in whole class discussions, and generally their interest began to shift from social to mathematical concerns.

When the students returned to their middle school classes in September we kept track of their achievement. Researchers visited the students' classes, to observe the teaching approaches they were experiencing and their participation. These observations worried us as we saw students sitting in rows, in silence, working through short, narrow questions requiring single procedures for solutions. Sadly, the students were returning to the same maths environments that most had told us they hated and students received limited or no opportunities to use the learning approaches emphasized in the summer. This did not give us much hope that the maths classes we had taught over the summer, in which students had been so engaged and excited to learn mathematics, would have a long term impact. It is very difficult for students to return to the same environment in which they have done badly before and do better, even after a summer of enjoying mathematics and learning new ways of working. This is why the home is such an important place for students to be encouraged to work in good ways. It was pleasing that the maths grades of our students did increase *significantly* in their next term at school, whereas the control group who attended the same summer school but did not attend our classes did not increase their maths grades significantly. Unfortunately, but not surprisingly, the students' renewed enthusiasm and increased achievement did not last, and their grades had fallen again by the following term.

Some of the students who attended our classes gained higher grades in the term following the summer and continued to achieve higher grades in maths. For these young people the relatively short maths intervention had given them a new approach to maths that they were able to continue to use. When we interviewed Lisa, one of the students who had maintained her improved achievement and asked her how the summer was helping her in the regular school year she said: 'When I don't know how to solve a problem the way the teacher does it, I have other ways to solve it.' Melissa, a student who had got an F before the summer and then achieved an A afterwards told interviewers that the most useful part of the summer had been learning strategies, in particular, "to ask a question if you're not sure" and to look for patterns. She also said "I used to hate maths and I thought it was boring but in this class maths was a lot of fun". Melissa's enjoyment and her learning of strategies certainly had a big impact on her achievement. I will describe the ways in which the strategies that we taught the students during the summer can be encouraged in the home or other classrooms in the next chapter.

The research results on the whole group of students we taught were very positive but the stories of individuals are probably most interesting and most insightful in understanding why many children are turned off maths. These are 4 of the students:

Jorge, who needed a chance and an opportunity[12]

Jorge came to our class with a history of D's and F's in his school maths classes, and school in general. He walked into our summer maths class on the first day with a huge smile on his face, joking with three of his friends. As on most days, he was dressed in baggy blue jeans and an Oakland Raiders cap that he only took off after much cajoling from his teacher. During that first lesson Jorge and his friends laughed and talked their way through the time, doing very little mathematics. Jorge was a social force in the classroom; funny and charming, he could pull his friends off task in an instant. Watching Jorge in class it was easy to see why he wasn't doing well in school. Jorge had all the markings of a "bad boy" who would try to get away with doing as little as possible in maths class and also keep other students from doing good work.

Observing Jorge during the later weeks of summer school it was still possible to catch glimpses of the "bad boy" behaviour that we saw on the first days. He still took a while to get started on tasks, still made jokes during discussions, and would still try to whisper something to throw off other students going to the board. However, he was also taking mathematics *extremely* seriously and, as he wrote in one of the surveys, he was working harder than he had *in any other maths class*. During discussions he listened to his classmates, and sometimes volunteered to present his own thinking. Over time, he presented more often and with fewer jokes to deflect attention away from his work. In both the whole class and in groups he more often talked with classmates seriously about mathematics, rather than always shifting conversations into the social territory with which he was more comfortable and in which he, an under-performing mathematics student, had more status. In a particularly striking example of Jorge's move to take maths more seriously, he and two other students spent over an hour during one lesson grappling with the generalization for a challenging pattern. They kept their focus only on the problem for almost this entire time, even moving to a different area of the room when two girls started working on one side of their table. Jorge did not just follow his peers' conversation; he also kept the other boys on task, asking questions when he was confused and volunteering ideas. He was deeply engaged in this task for a long period of time, a striking turnaround from the first day when he seemed to go out of his way to avoid serious mathematical work.

Jorge's comments in journals and interviews reveal aspects of the summer environment that may have contributed to his willingness to work more seriously in the class (and, importantly for such a "cool" kid, to be *seen* working more seriously). Notably Jorge says that he worked *harder* in the summer school class than his regular classes, although he also describes the summer school classes as "funner". He explains that he works harder in summer school because:

'In our [regular] class they give us, like, easy problems and all that. And in this class you give us hard problems to figure out. You have to figure out the pattern and all that.'

Asked what advice he would give maths teachers to help them teach better, he says that he would tell them to "give harder problems." These comments suggested to us that he appreciated being taken seriously as a capable mathematics student.

When Jorge talked about the hard problems he was given in the summer school class, he talked both about being able to "figure out" the solutions himself and about being given the *time* to do that. He explained that he enjoyed working on pattern problems because:

> 'We stay on it longer . . . so we can really get to know how to do pattern blocks and everything, and try to figure out the pattern.'

Jorge also talked about the value of working in groups. Another piece of advice he said he would give to teachers was to put students in groups, because:

> 'You learn more from other people's ideas.'

On the days that Jorge became deeply engaged in mathematics, he was working on problems that could be viewed in many different ways; and he was working with high attaining boys whom he respected. Jorge left our classes proud of his maths work and ready to work hard in his regular maths classes, but he returned to a situation in which he had to work alone on problems that Jorge would not have considered "hard" or interesting. Jorge quickly reverted to socializing instead of doing maths work. In the first term after the summer he received a D, in the next term he received an F.

Rebecca, who needed to understand.

Over the twenty or so years I have been a school teacher and researcher I have met many students like Rebecca – and usually they are girls. Rebecca was conscientious, motivated and smart, and even though she attained A+ grades in mathematics she did not feel that she was good at it. Rebecca, like many other students, could follow the methods the teachers demonstrated in her regular classes, and reproduce them perfectly, but she wanted to understand mathematics and she did not feel that the procedural presentations of mathematics she experienced gave her access to understanding. Rebecca described her previous maths classes as always the same – the teacher would start with a warm-up and then give students worksheets to go through individually. There were no classroom discussions and the questions they worked through were shallow and procedural. In interviews Rebecca and her friend Alice were asked if they regarded themselves as "maths people". When Rebecca said no Alice protested saying that Rebecca had won an award in maths. I asked Rebecca about this and why she did not regard herself as a maths person. She said it was because:

'I can't remember things well and there is so much to remember.'

When students tell me that there is a lot to remember in mathematics I know that they are being taught badly and the subject is being misrepresented. I know also that they are being overwhelmed by the procedures teachers show and have come to believe that they must memorize them all, instead of understanding the concepts that link the procedures and render memorization unnecessary. Rebecca explained to me that you "have to remember" in maths and that it was hard to remember because "you don't use it in life". All of the procedural lessons she experienced made Rebecca feel a failure even though she gained A+'s. Our maths classes were very different as we taught students the strategies and ideas that connect mathematical procedures and give access to understanding. The notes that teachers had written on Rebecca in their recommendations for summer school told us that she would not, under any circumstances, talk in class as she was painfully shy. But as Rebecca came to enjoy the maths problems we presented and understand the mathematics we taught she not only talked in class but even chose to go to the board and show her work to the whole class. We were surprised and thrilled to see Rebecca participate so publicly and knew that her participation spoke to the mathematical understanding she was gaining for the first time.

Rebecca told us that she most appreciated the summer classes because they allowed her to understand mathematics. When she was asked about the differences between the summer class and previous maths classes she said:

'We stretch the problems a lot more. Before we would just get answers and not really stretch them. (. . .). Like patterns you can really look at it and find how it grows. (. . .) Stretching the problems helps you understand the problem more. In this class you don't stop when you get the answer, you keep going'.

Rebecca was clear that working on longer problems that she was able to explore and extend gave her access to an understanding she had never had before. Some readers may worry that students who are high achievers cannot learn in the same classes as students who struggle with maths, but Rebecca did not give any indication, in class or in interviews that she was held back in such a mixed setting. On the contrary, she reflected in her journal that the summer school class:

'Has been more useful because we take the time to make sure everybody understands everything and we use different methods of learning.'

When Rebecca was asked what she had learned she talked about many aspects of the class. She said that she had learned to generalize algebraic patterns "instead of just staring at them", that she had learned to multiply double digit numbers in her head (from the number talks), that she had learned strategies such as organization, and question asking and, in her words, she had learned:

'To think beyond the answer to the problem.'

Rebecca summarized her experience saying:

'I enjoy this class more than any other maths class I have had because we are learning maths in a fun way, and I think we can learn as much or more maths this way as in a textbook'.

Rebecca came to our maths class with a history of A+'s and she continued to get A+'s in the year afterwards, but we hoped that she no longer believed mathematics to be a subject that had to be remembered, without understanding or enjoyment.

Alonzo, who needed to use his ideas[13].

Alonzo could easily have been mistaken for a student who was overly concerned with his own popularity. He was always surrounded by other students, before and after class, although he seemed as interested in being adored by them as he was in observing them from just outside the crowd. Alonzo could be described as the "strong silent type" because of his striking, tall athletic build and quiet, fiercely observant ways. For the first few days Alonzo would slip into class undetected and pull the brim of his baseball cap down, as if hiding, silently watching the activities unfold before him. As the summer progressed Alonzo's behaviour changed and we came to realize that the mathematics he was working on was allowing Alonzo's curiosity and creativity to take root and bloom. As Alonzo experienced the opportunity to use his ideas in maths he started to become a different kind of student in class.

Like many other students who attended summer school, Alonzo gained an F in his last maths class and was forced to enroll in the summer school class by his maths teacher (this is typical in the US as summer school classes are seen as the chance to catch up with school subjects). Alonzo described his previous class as one where the teacher talked and the students plodded through worksheets, and where group-based collaborations were extremely rare. Maths learning in that class was largely a repetitious, worksheet-based experience. As Alonzo described it:

'In regular maths class we had to go to work. We couldn't talk or we couldn't like [offer], "oh, can I help?" No. We had to like, he would just give us a paper, a pencil and put us to work.'

In informal conversations during class, Alonzo described his previous class as boring and frustrating.

One of our activities in which Alonzo's curiosity and creativity was piqued was called "Staircases." In this task students were asked to determine the total number of blocks in a staircase that grew incrementally from 1-block high, to 2-blocks high, to 3-blocks high, and so on, as a move toward predicting a 10-block high staircase, a 100-block high staircase and finally algebraically expressing the total blocks in any staircase. Students were provided with a box full of linking cubes to build the staircases if they wished.

A 4-block-tall staircase; total blocks = 4 + 3 + 2 + 1 = 10

Midway through the time allotted for the activity, Alonzo seemed to be playing with the linking cubes and not working on the problem. Drawing near, we saw that Alonzo had decided to modify his staircase so that it extended in four directions. Thus a 1-block high staircase had a total of five blocks, a 2-block high staircase had a total of fourteen blocks and so on.

Alonzo's staircase 5 blocks   *5 + 9 = 14 blocks*   *5 + 9 + 13 = 27 blocks*

What had looked like messing around was in fact Alonzo's creativity and curiosity conspiring to create a problem that was more diagrammatically and algebraically involved than the one that had been originally presented to him. Eventually other students began looking in on Alonzo's work and, after convincing themselves they had mastered the original problem, tried to do the "Alonzo staircase" problem.

The teacher of Alonzo's class was so impressed with his innovative approach to the problem, as well as his growing interest in class and willingness to push himself to work at high levels, that she decided to phone home and tell his parents of his achievements. During this call his mother described Alonzo as a junior engineer who had done several small projects around the house – among them was a mechanical pulley-system that Alonzo designed using dental floss and a stack of pennies. He used this in his bedroom to turn the light switch off without getting out of bed. Alonzo's mother described how he locked himself in his room while working on the project, coming out only to gather additional materials. In recounting the project to his mother, Alonzo said he'd had to experiment with different sized stacks of pennies until he had figured out the exact weight it took to turn off the light switch without breaking the dental floss. Despite his creativity and apparent interest, she explained that she hadn't heard a single good word about Alonzo's mathematical ability since he was in year 4. Alonzo's mother was very grateful for the phone-call and asked if it would be possible for us to keep in touch with her after the summer to help lobby his teachers for more opportunities in which Alonzo could explore mathematics and use his creativity and resourcefulness. As we watched Alonzo work and reflected on the achievements his mother described it was difficult to understand how he could ever achieve F grades in maths classes.

Over the course of the five weeks, the open, problem-solving, group-based format of the summer program not only allowed Alonzo to play out his mathematical curiosity, but also encouraged him to take a more visible role in the classroom. During an activity called "Cowpens & Bullpens" students had to determine how many lengths of fencing were required to contain an increasing number of cows given certain fencing parameters. At the close of the activity student volunteers were asked to share their solutions and Alonzo was the first to volunteer. He strode to the front of the room and carefully diagrammed his fencing strategy on the overhead projector, documenting his work numerically and eventually, algebraically.

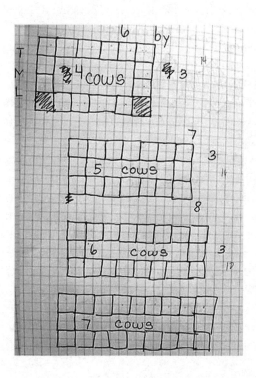

In that moment the young man who once hid behind his baseball cap had all but disappeared and, in his place, a seemingly more confident maths student willing to share his ideas with the entire class, had emerged. Alonzo was one of the highest scorers on the algebra test we gave at the end of the summer, with an impressive 80%, some 30% higher than when he took the test in the school year. But when Alonzo returned to his school classes and was required to work on short questions in silence he again withdrew and the grade he received for the class was, again, an F.

Tanya, who needed discussion and variety.

Tanya described herself as a 'people person', a description that was easy to understand as she spent much of her time in maths class talking to others. Tanya and her friend Ixchelle would talk and laugh as they worked and often whisper conspiratorially. Tanya did not seem to be aware that teachers were noticing her behaviour and they would frequently ask her to be quiet. Tanya's talkativeness must have concerned many of her teachers, indeed her previous maths teacher recommended she attend summer school and among her comments wrote, "Tanya has a voice that carries easily." This not-so-subtle reference to Tanya's assertive, boisterous personality implied that her exuberance and social disposition as a student needed controlling. But Tanya's own perspective on her social needs was interesting. In interviews Tanya described herself as needing an environment where students could work together because, as she explained:

'Then I know if I'm doing wrong. If my way isn't the only way, if my way can be easily comprehended, or if it's a little too hard so I should try something different.'

But Tanya's previous maths classes did not allow collaborative work, as Tanya described to us:

'For the last past year, maths was the hardest because you're not supposed to talk, you're not supposed to communicate . . . [In other subjects] they let you talk (. . .) In maths, in maths class, [the teacher says] just like 'Okay, get to work, no talking, be quiet, shhhhh!''

Tanya went on to describe the class and the teacher as dominated by silence:

'[it is] a whole hour of silence. That's a good class to [the teacher].'

Our summer classes were very different and for Tanya, and many other students, the collaboration they experienced in class was critical to their engagement. Tanya spoke eloquently about the opportunities afforded by the discussions, and her reasons for valuing them were not to do with socializing or enjoying her time but understanding mathematics, for example:

'You can do it multiple ways . . . It's like the way, the way our schools [in the school year] did it is like very black and white, and the way the people do it here [in summer school] it's like very colourful, very bright. You have very different varieties you're looking at; you can look at it one way, turn your head and all of a sudden you see a whole different picture.'

Tanya enrolled in summer school because she was not doing well in her previous maths class. In our class Tanya did extremely well and attained one of the highest scores in the class on the final algebra test. In interviews Tanya reported that she:

'learned more in five weeks than in almost a year.'

This statement was probably exaggerated but it did reflect Tanya's appreciation of the learning opportunities she received. Tanya described the summer environment as highly communicative, fun and a class in which she learned a lot. When asked what aspects of summer school she would recommend for teachers in the regular school year, Tanya advised:

'More communication . . . a lot of communication, a lot of feedback from teachers and from students to teachers.'

When asked to reflect on her performance in the summer maths class Tanya exclaimed:

'Very different! Normally [I am] the lowest one in the class, always needing help'.

One of the key features of the program that Tanya seemed to benefit from was the varied nature of the mathematical tasks and the ways in which they encouraged multiple methods and approaches. Tanya noted that she especially appreciated the tasks that included apparatus (blocks, tiles, linking cubes), the pattern work, whole-class activities such as the number talks, group and pair work, journal writing and in-class presentations. In interviews she remarked:

'I've enjoyed finding the patterns and stuff like that and groups . . . and it really helped me, the number talk really helped me to do stuff mentally.'

Tanya seemed to revel in the plurality of ways she had to engage in maths and described the tasks as more challenging, interesting, accessible and fun:

'It was much funner. You not only got to like just see the problem, you got to think it, you got different parts . . . you got to smell it, you got to eat it. And then after you finish the task that's given to you, you need to have another assignment to be like, well what if this changed, and you did this with that, instead of, you know, that. It just, it just opens your mind, and makes it harder, a new way of thinking.'

Through interviews in the summer and the following term, as well as in-class observations, Tanya revealed how the open, group-based format of summer school allowed her to express herself socially *and* develop mathematically. Tanya, like many of our summer school students, wanted desperately to appreciate maths and her experience of it on her own terms. She was a 'people person' who strove to see the colour in maths and yet found herself in situations which were monotonous and grey. Tanya was not a student who had been turned off maths (yet), but she clearly wanted something different. Our course gave Tanya a different way to think about and understand maths. Tanya's positive experiences in the summer had an impact on her, and she tried very hard to engage with her regular maths class in the fall, receiving a very respectable B grade for maths, but by the second quarter her attainment had dropped to a D again. Tanya's low attainment, which did not seem to fit her potential or enthusiasm from the summer, can probably be explained by the silent and monotonous nature of her school maths class, as she aptly described:

'I would say . . . the only way to describe summer school is very colourful and then this [class in the school year] is just still, ugghhh, black and white. And you just wanna ask 'Can I have a little bit of yellow?'

Tanya's request, and that of so many students – to experience a mathematics that is varied and interesting, is eminently reasonable.

In our summer school we gave students activities to work on and strategies to use, that I will share in the next chapter. We taught the students that mathematics was a subject with many different methods and that numbers could be used flexibly. We encouraged the students to have confidence in their own mathematical ability, something that no doubt contributed to their

8 / Giving Children the Best Mathematical Start in Life

Activities and Advice.

Biographies of the lives of mathematicians, such as Sofia Kovalevskaya, Lipman Bers and William P. Thurston, are fascinating to read, but I am always struck by the fact that so many were inspired not by school teaching, but by interesting problems or puzzles that were given to them by family members in their homes.[1] I was one of the people given the greatest mathematical start in life because my Mum brought home puzzles and shapes for me to play with as a young child. It was many years later, at age 16, that I was also inspired by a great maths teacher, who asked her students to talk about mathematics, which gave me access to a deeper understanding than I had had before. It is important not to underestimate the role of simple interactions in the home, and the role of puzzles, games, and patterns, in the mathematical development and inspiration of young people. Such problems and puzzles can be more important than all of the short questions that children work through in maths classes. Sarah Flannery, the young woman who won the European Young Scientist of the Year in 1999 for her mathematical work reflects upon the fact that as a child she was given puzzles to work on at home, and how these were more important in her own mathematical development than the years of maths that she was taught in classes. Mathematical puzzles and settings are the ideal way for parents – or teachers – to encourage their children into maths and this chapter will outline some of the ways that these and key ways of working in maths can be introduced to children with great effect.

Part 1: Settings, Puzzles and Questions.

1.1 *Mathematical Settings.*

All children start life being excited by mathematics and parents can become a wonderful resource for the encouragement of their thinking. Mathematical ideas that may seem obvious to us – such as the fact that you can count a set of objects, move them around and then count them again and you get the same number – are fascinating to young children. If you give children, of any ages, a set of pattern blocks, or Cuisenaire rods and just watch them you will see them do all sorts of mathematical things, such as ordering the rods, putting them into shapes, and making repeating patterns. At these times parents need to be around to marvel with their children, to encourage their thinking and to give them other challenges. One of the very best things that parents can do to develop their children's mathematical interest, is to provide mathematical settings and to explore mathematical patterns and ideas with them.

There are many books filled with great mathematical problems for children to do, but it is my belief that the best sort of encouragement that can be given at home does not involve sitting children down and giving them extra maths work, or even buying them mathematical books to work on, it is about providing settings in which children's own mathematical ideas and questions can emerge and in which children's mathematical thinking is validated and encouraged. Fortunately mathematics is a subject that is ideally suited to the provision of interesting settings that can encourage this. One of my Stanford doctoral students, Nick Fiori, taught a number of mathematics classes in which he provided students with different mathematical settings – some photographs of the settings are given on the right of the page – and encouraged students to come up with their own mathematics questions within these settings. Students of different ages and backgrounds, including those who

Coloured beads of different shapes, and strings.

Nuts, bolts, washers, and coloured tape.

Cards from the game SET®.

Dice of various colours.

'Snap cubes' of various colours.

A *square lattice grid with coloured pegs.*

had suffered very negative experiences with maths in the past, set about posing important questions. In some cases these questions led to maths problems that were completely new and had never been solved before. Fiori documents the various mathematical ways that students worked, and he makes a very good case for the encouragement of similar problem posing in school classrooms, for at least some of the time.[2] I agree with him, and would also encourage parents to provide such settings in the home, for children of all ages.

Children's play with building blocks in the early years has been identified as one of the key reasons for success in mathematics all through school.[3] Indeed the fact that boys are usually encouraged to play with building blocks more than girls is the reason that differences often occur in spatial ability among boys and girls, which impacts mathematics performance greatly. Any sorts of play with building blocks, interlocking cubes, or kits for making objects, is fantastically helpful in the development of spatial reasoning, which is fundamental to mathematical understanding.

Geoboards with rubber bands.

In addition to building blocks, other puzzles that encourage spatial awareness include jigsaw puzzles, tangrams, rubik cubes and anything else that involves moving objects around, fitting objects together or rotating objects. Mathematical settings need not be sets of objects, they can be simple arrangements of patterns and numbers in the world around us. If you take a walk with your child you will stumble upon all sorts of places that can be mathematically interesting, from house numbers to gate posts. The creative mind at work will see mathematical questions and discussion points everywhere – there is always something mathematical that can be brought into focus, if we only remember that that is what we should be doing.

'Pattern blocks' of various colours and regular shapes.

Sticks of different unit lengths with eyelets and string.

Measuring cups with simple fraction volumes and a bowl of water.

Pinecones of various shapes and sizes.

Eleanor Duckworth, a professor of education at Harvard University, is the author of an essay called 'The Having of Wonderful Ideas'[4] which makes an extremely important point – the most valuable learning experiences children can have come from their own thoughts and ideas. In Duckworth's essay she recalls an interview she had with some 7 year old children in which she asked them to put 10 drinking straws, cut into different lengths, in order, from the smallest to the biggest. When Kevin walked into the room he announced "I know what I'm going to do" before Duckworth had explained the task. He then proceeded to order the straws on his own. Duckworth writes that Kevin didn't mean ' "I know what you're going to ask me to do." He meant "I have a wonderful idea about what to do with these straws. You'll be surprised by my wonderful idea."' Duckworth describes Kevin working hard to order the straws and then being extremely pleased with himself when he managed it. For Kevin the experience of ordering straws was much more worthwhile because he was working on his own idea, instead of following an instruction. Research on learning tells us that when children are working on their own ideas, their work is enriched with cognitive complexity and enhanced by greater motivation[5,6,7]. Duckworth proposes that having wonderful ideas is the 'essence of intellectual development' and that the very best teaching that a parent or teacher can do is to provide settings in which children have their most wonderful ideas. All children start their lives motivated to come up with their own ideas – about mathematics and other things – and one of the most important things a parent can do is to nurture this motivation. This may take extra work in a subject like mathematics, in which children are wrongly led to believe that all of the ideas have been 'had' and their job is simply to receive them,[8,9] but this makes the task even more important.

1.2 Puzzles and Problems.

In addition to the provision of interesting settings, another valuable way to encourage mathematical thinking is to give children interesting puzzles to work on. Sarah Flannery, the young woman who became the European young scientist of the year for inventing a 'breathtaking algorithm' has written a fascinating book called: *In Code: A Mathematical Journey.*[10] Flannery's book describes her mathematical development and it is a very useful resource for parents who would like to encourage the best mathematical start in life, and I don't think that parents need to be the mathematics professor that her father was in order to be successful, they just need the enthusiasm. Sarah Flannery talks about the way her mathematical development was encouraged by working on puzzles in her home. She said that although she and her siblings much preferred outdoor sports her father would give them intriguing puzzles to think about in the evenings and these captured their young minds. Because her father was a mathematics professor and because she was so good at mathematics, people have often assumed that she was given extra maths help at home, but Flannery shares something very important with her readers. She says: 'Strictly speaking, it is not true to say that I or my brothers don't get any help with maths. We're not forced to take extra classes, or endure gruelling sessions at the kitchen table, but almost without our knowing we've been getting help since we were very young – out-of-the-ordinary help of a subtle and playful kind which I think has made us self confident in problem-solving. Ever since I can remember, my father has given us little problems and puzzles. I have often heard, and still hear, "Dad give us a puzzle." These puzzles challenged us and encouraged our curiosity, and many of them made maths interesting and tangible. More fundamentally they taught us how to reason and think for ourselves. This is how puzzles have been far more beneficial to me than years of learning formulae and "proofs."[11]

Sarah Flannery gives examples of the sorts of maths problems she worked on as a child that caused her to be so good at mathematics, here are three of my favourites:

The Two Jars Puzzle: Given a five-litre jar and a three-litre jar and an unlimited supply of water, how do you measure out four litres exactly?

The Rabbit Puzzle: A rabbit falls into a dry well, thirty metres deep. Since being at the bottom of a well was not her original plan, she decides to climb out. When she attempts to do so she finds that after going up three metres (and this is the sad part) she slips back two. Frustrated, she stops where she is for that day and resumes her efforts the following morning – with the same result. How many days does it take her to get out of the well?

The Buddhist Monk Puzzle: One morning, exactly at sunrise, a Buddhist monk leaves his temple and begins to climb a tall mountain. The narrow path, no more than a foot or two wide, spiralled around the mountain to a glittering temple at the summit. The monk ascended the path at varying rates of speed, stopping many times along the way to rest and eat the dried fruit he carried with him. He reached the temple shortly before sunset. After several days of fasting he began his journey back along the same path, starting at sunrise and again walking at variable speeds with many pauses along the way, finally arriving at the lower temple just before sunset. Prove that there is a spot along the path that the monk will occupy on both trips at precisely the same time of day.[12]

Flannery talks about the ways these puzzles encouraged her mathematical mind because they taught her to *think* and *reason*, two of the most important mathematical acts. When children work on puzzles such as these they are having to make sense of situations, they are using shapes and numbers to solve problems and they are thinking logically, all of which are critical ways of working in mathematics. Flannery reports that she and her siblings would work on problems that her father had set for them each night over dinner. I like the sound of this ritual although I also appreciate that it is a difficult one to achieve in a busy household at the end of a tiring day. And puzzles do not have to be set by parents for children to work on, they can be something that children and parents work on together, each week, month or more occasionally. A good friend of mine called Cathy, told me that she used to put a maths puzzle into her son's lunchbox every day, all the years he was growing up. Her son is now a brilliant mathematician. Whether a daily ritual or something less frequent, puzzles are incredibly useful in mathematical development, especially if children are encouraged to talk through their thinking and someone is there to encourage their logical reasoning. If children get in the habit of applying logic to problems and to persisting with problems until they solve them, they will learn extremely valuable lessons for learning and for life.

Some recommended websites and books of mathematical puzzles are given at the end of this chapter.

1.3 *Asking Questions.*

When exploring mathematical ideas with young people it is always good to ask lots of questions. If your child is young enough she or he will like giving answers to questions and it will help them develop a mathematical mind. Good questions are those that give you access to your child's mathematical thoughts, as these will allow you to support their development. When I am called over to students who are stuck in maths classes I almost always start with: 'what do you think you should do?' and if I can persuade them to offer any ideas I will ask 'why do you think that?' or 'how did you get that?' Often children who have been taught traditionally will think they are doing something wrong at this point, and quickly change their answer, but over time my students get used to the idea that I am interested in their thinking, and I will ask the same questions whether they are right or wrong. When students explain their thinking I am able to help them move forward productively at the same time as helping them know that maths is a subject that makes sense and that they can *reason* their way through mathematics problems.

Pat Kenschaft, an American mathematician, has written a really useful book for parents called *Math Power: How to help your child love maths even if you don't* in which she quotes a Swarthmore professor called Heinrich Brinkmann. This particular professor was known on the Swarthmore campus for being able to find something right in what every student said. No matter how outrageous a student's contribution or question, he could respond, "Oh I see what you are thinking. You're looking at it as if . . ."[13] This is a very important act in mathematics teaching because it is true that unless a child has taken a wild guess, then there will be some sense in what they are thinking – the role of the teacher is to find out what it is that makes sense and build from there. A parent in the home can do the sort of careful inquiring and guiding as they help children with mathematics that it is hard for teachers in a classroom of 30 or more children to do. If a child gives an answer and just hears that it is wrong, they are likely to be disheartened, but if they hear that their thinking is correct in some ways and they learn about the ways that it may be improved, they will gain confidence, which is critical to maths success.

Children should also be encouraged to ask questions, of themselves and others. I once worked with an inspirational teacher, called Carlos Cabana, who, when asked for help, would prompt students to pose a really specific question that he and they could think about. As the students phrased their question specifically, they would be able to see the mathematics more clearly and they would often be able to answer the question themselves! Another great teacher I worked with, called Cathy Humphreys[14], always says that she never asks a question that she knows the answer to. What she means by this is that she always asks students about their mathematical methods and reasons, which she could never know in advance. These are the most valuable questions for any maths teacher to ask, as they give the teacher access to students' developing mathematical ideas. Pat Kenschaft puts it well: 'If you can tap into the real thoughts of the person before you, you can untangle the knots around their mathematical inner light.'[15] The ideal way in which to tap into a learner's mathematical thoughts and to discover their 'mathematical inner light' is to provide interesting settings and problems, to gently probe and question, and to encourage their thinking and reasoning.

When you are working with your child on maths it is important to be as enthusiastic as possible about maths. This is hard if you have had bad mathematical experiences, but it is very important. Parents, especially mothers of girls, should never, ever say I was hopeless at maths! Research tells us that this is a very damaging message, especially for young girls[16]. Pat Kenschaft, talks about being joyful with maths. As she says: 'make your mathematical interactions with your own child as much fun as you can, especially when your child is young. Fortunately, the time when a child is most vulnerable is also the time when you have least reason to be anxious about your own inadequacies. You can count. Count and laugh. Count and sing. Count and dance.[17]' She makes an important point – there is no reason for any parent to be negative about the mathematics of early childhood as even the most mathsphobic of parents would not have had negative experiences of maths before school started. And the birth of your own children could be the perfect opportunity to start all over again with mathematics, without the people who terrorized you the first time around. I know a number of people who were traumatised by maths in school but when they started learning it again as adults, they found it enjoyable and accessible. Parents of young children could make maths an adult project, learning with their children. Parents regularly read to children but rarely discuss maths with them – yet this can be so valuable for children's development and can be a great project to work on together.

Mathematical conversations should be relaxed and free from pressure. Fear and pressure impede learning and children should always feel comfortable when offering their ideas in maths. Parents and teachers should never appear irritated or judgmental if children make mistakes. Maths, more than any other subject, can cause panic, which stops the mind working[18]. I usually start any work with children by telling them that I love errors because they are really good for learning. I mean this because it is through making errors that children learn the most – as mistakes give them a chance to consider, revise and learn new things. When I am working with children and they say something that is incorrect, I consider their thinking with them, and see this as an important opportunity for learning. When students know that I am not judging them harshly and that I genuinely value errors they are able to think more productively and learn more.

In addition to the settings, puzzles, and mathematical conversations that parents can provide and stimulate at home, to give children the best mathematical start in life, there are particular ways of working in maths that are known to be critical to mathematical success. These can also be taught to children in the home, or in school, as I will explain now.

Part 2. Mathematical Ways of Working.

2.1 Strategies for Solving Problems.

Many people have studied the work of experts in various fields such as chess, basketball and physics and set out what they actually do. Mathematics is no exception and researchers have studied the general strategies used, time and time again, by successful mathematical problem solvers. One of the most widely recognized and accepted records of the ways mathematicians solve problems came from a Hungarian mathematician called George Pólya. In 1957 he wrote a book called 'How to Solve It'[19] in which he set out a list of strategies that successful problem solvers use. Pólya's list has received a lot of attention and support from across the world and is still one of the most accepted around. Pólya said that when experts solve mathematics problems they first do work to Understand the Problem. They ask questions of themselves such as – what is involved in this problem? What are the questions to be answered? In the next stage they Make a Plan. At this stage, mathematicians engage in some very important actions, often missed by low achieving students such as:

Drawing the problem
Making a chart with the numbers
Trying a smaller case

For example, if you are asked how many different ways there are to shade 2 squares in this shape:

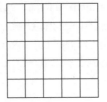

Then it is a good idea to try it with this case first:

In the third stage they carry out their plan, checking each step. And in the fourth stage they look back at their work, thinking about whether their answers are reasonable. Some of these general strategies may seem obvious to readers but they are often missing from the work of low achievers in maths. One of the biggest mistakes students make with maths problems is that they often rush in and do *something* with the numbers, without really considering what is being asked of them,[20] whereas successful problem solvers spend some time really thinking about the problem – considering, what is being asked here? Successful problem solvers then do some very mathematical things such as drawing the problem, making a chart or trying a smaller case, that unsuccessful students don't think to do. In mathematics I find drawing absolutely critical and whenever I see a problem, such as the three given above, the first thing I do is draw the situation presented, as best I can. In fact I would usually be hopelessly lost if I did not stop to draw problems. What this does for me, and for many other problem solvers, is help me see what is what, and how things are related to each other. When I work with school children and they ask me for help I frequently suggest that they draw what they are thinking about, and invariably they find this extremely helpful. The importance of the other two strategies that mathematicians use, of trying a smaller case and making a chart, were recently brought home to me vividly as I will describe shortly.

I will illustrate the importance of these 2 strategies with the chessboard problem. The main task of this problem is to work out how many squares there are on a chessboard – and the answer is not 64. What makes this problem difficult is the fact that there are many different sized squares on the board, from 1 × 1 squares, which are the smallest, to 2 × 2 squares all the way up to an 8 × 8 square, which is the whole board.

For example, this is a chessboard showing a 2 × 2 square and a 4 × 4 square:

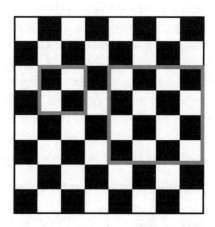

On the first day of the summer school teaching that I described in the last chapter, with our mixed group of low and high achieving 7th and 8th graders, we presented the students with this problem and we asked the students to generalize and to tell us a way of finding out how many squares there would be on a board of any size. One of the features of this problem that makes it challenging, especially in the initial recording stages, is that the squares overlap.

So students need to be very careful when they count squares and very systematic in their recording. When we set this problem an interesting thing happened – the high achieving students thought about the problem and decided it was asking them about different sized squares. They then started to count systematically, marking their drawings of boards so that they could keep a record of those they had counted. An example of the strategy Ella used to keep track of the 2 × 2 squares is shown below:

Stage 1. Marking the center of the 2 3 2 squares with a dot:

Stage 2. Finding all of the 2 3 2 squares:

The students then recorded into charts how many different sized squares there were, Ella's chart is shown below:

Size	how many	
1 × 1	~~64~~ 64	₄ 64
2 × 2	49	49
3 × 3	(6×6) = 36	36
4 × 4	(5×5) = 25	25
5 × 5	(4×4) = 16	16
6 × 6	(3×3) = 9	9
7 × 7	(2×2) = 4	4
8 × 8	①	+ 1
		204

64 + 49 + 36 + 25 + 16
+ 9 + 4 + 1 = 204 squares

But the low achieving students did something very different. They also realized that different sized squares were needed but they started counting them in a disorganized way, and did not get them all. They then told us numbers of squares, without having counted systematically or organized their numbers in a chart. As the days and then the weeks went by it became clear that one of the main things that was hurting the performance of the low achievers, in many features of their work, was their lack of careful recording or organization of ideas. They often did not stop to consider problems at the beginning, and they did not organize their thinking. So they would normally rush in and try anything with the numbers, often getting hopelessly lost. We worked with the students on this, teaching them to do such things as draw charts and tables, and to be systematic in their thinking. Interestingly, when the students became more careful about being organized – drawing charts and tables to help them, they became much more successful on a range of maths problems.

The other strategy that Pólya and others highlight, of taking a smaller case, also proved to be a major barrier for the low achieving students, more so than I could have imagined. The importance of this strategy can also be seen in the case of the chessboard problem. When we asked the students to generalize, and to think about the numbers of squares in an 8 × 8 square, we were asking them to think about patterns. A successful problem solver, when faced with this question, would see that there are different numbers of squares on different sized boards, so that the number of 2 × 2 squares, or 3 × 3 squares grows every time you have a bigger board. We were asking the students to say how those patterns grew. In the case of the 8 × 8 board there were lots of patterns to look at, how many 2 × 2 squares are there? How many 3 × 3 squares? 4 × 4 squares etc? A very good and mathematical strategy at this point is to look at the squares in a smaller, more manageable case first. For example to take a 2 × 2 mini chessboard and a 3 × 3 board and think about how many squares there are in each. The low achieving children in our class did not use this strategy. Not only did they not use it, they resisted it *very strongly* when we asked them to use it. In the last chapter I talked about the importance of viewing and using numbers flexibly and seeing numbers as something that can be taken apart and put back together again. This sort of flexible outlook is needed in all mathematics, including the problems that are given. Successful problem solvers in mathematics know that if asked to find the numbers of squares in an 8 × 8 board, it is a good idea to find the numbers of squares in a smaller sized board first. But the low achieving students would not do this, somehow seeing it as cheating or breaking some silent rule. Over the weeks we experienced many students finding it extremely difficult to try a smaller case, which is when I realized that this was a really hard lesson for them, as it went against their *ladder of rules* view of mathematics that I talked about in the last chapter. All these students had ever done in maths classes, was work on sets of problems that they had been given, rehearsing a method over and over again, and here we were asking students to *change* the problems before they worked on them. Not surprisingly this seemed like a very strange action for them, but without doing this many of the maths problems that students were given became close to impossible. Over the weeks we worked with the students on the strategy of trying a smaller case, and as they gradually began to see mathematics as flexible, and to use numbers flexibly, they also began to use this and other important strategies.

The four stages of Pólya's cycle, of understanding the problem, making a plan, carrying out the plan and looking back were all neglected or missing in the work of the low achieving students, who would more typically rush into answering problems, without planning systematically, neglect to use key strategies and finish when they found an answer, without stopping to consider whether their answer was reasonable. Recently the US Department of Education in the US called together a group of mathematicians and mathematics educators to outline the future most important directions for research. I was one of the people selected to work on this task and we met, as a group, on a number of occasions. After many interesting and considered discussions the group decided that some of the most important actions in which mathematicians engage, are a set of what we termed 'mathematical practices'. These were defined as 'the mathematical know-how, beyond content knowledge, that characterizes expertise in learning and using mathematics. The term "practices" refers to specific things that successful mathematics learners and users do. Justifying claims, using symbolic notation efficiently, and making generalizations are examples of mathematical practices.'[21]. Interestingly the group also agreed that these practices, often neglected in classrooms, yet characteristic of high-level performance, are what separate high and low achievers in classrooms. It seems the high achievers have learned to work in these ways, whereas the low achievers have not. Problem solving strategies and mathematical practices are critical to teach children.

The very best way to teach children helpful strategies is to provide interesting settings, problems and puzzles that require the strategies and then to share and discuss successful methods and strategies, at frequent intervals. It can also be valuable to teach students mathematical strategies, more directly – and over a short space of time, as we did in the summer school teaching described in the last chapter. One researcher, called Kil Lee, used Polya's strategies with a group of 5th grade students in a short teaching intervention[22]. In Lee's experiment the instructors spent 5 class sessions emphasizing the strategies and demonstrating their usefulness in problem solving, then 15 sessions encouraging the students to use the strategies themselves in solving problems. When Lee compared the students who had received this intervention with those who had not, he found that the students who were taught strategies were less likely to jump into problems and do something with numbers before considering what was being asked, and more likely to draw the situations, make charts and consider special cases. This increased the students' success in mathematics classes and the students who learned the strategies were more successful than the control groups in solving maths word problems, even some weeks later. This research study showed that students as young as 5th grade could be taught important mathematical strategies and that it greatly enhanced their performance in school, on standard problems of the type that are typically used in mathematics classrooms in the US and UK[23].

2.2 *Number Flexibility*

An important mission for all parents and teachers is to steer children away from *the mathematical ladder of rules* that I discussed in the last chapter. Gray and Tall's research study[24] showed that successful students were those that used numbers flexibly, decomposing and recomposing them. This isn't difficult to do but it involves children knowing that that is what they *should* be doing. Fortunately there are specific and enjoyable ways of encouraging this number flexibility, that can be used with children of all ages. One of the very best methods I know for encouraging number flexibility is that of *number talks*, the activities I discussed in the last chapter. The aim of number talks is to get children to think of all the different ways that numbers can be calculated, decomposing and recomposing as they work. For example, you could ask a child to work out 17×5 in their heads without the use of pen and paper. This is a problem that looks difficult but becomes much easier when the numbers are flexibly moved around. So, for example, with 17×5 I could work out 15×5. I can do this in my head more easily as 10×5 is 50 and 5×5 is 25 giving me 75. I then need to remember to add 10 as I only worked out 15×5 and I need 2 more 5's. This gives me my answer of 85. Another way of solving the problem is not to work out 17×5 but to work out 17×10 which is 170 and then halve it. Half of 100 is 50 and half of 70 is 35 so I would get 85. As people work on problems like these, they become used to using numbers flexibly as well as generally sharpening their mental maths skills. The problem I would set children when working on number talks, is to find as many ways as possible of working out the answers. Most children will find this challenging and fun.

One of the great advantages of 'number talk' problems is that they can be posed at all sorts of levels of difficulty, and there is an endless selection of possible problems, so they can be fun for children and adults of all ages. Here are some starters, which are at different levels of difficulty:

Addition/Subtraction	Multiplication
$25 + 35$	21×3
$17 + 55$	14×5
$23 - 15$	13×5
$48 - 17$	14×15
$56 - 19$	17×15

Some good prompts to use while you are working with children are:

- How did you think about the problem?
- What was the first step?
- What did you do next?
- Why did you do it that way?
- Can you think of a different way to do the problem?
- How do the two ways relate?
- What could you change about the problem to make it easier or simpler?

Number talks are an excellent way to teach children, of any age, to decompose and recompose numbers, which is extremely valuable in their mathematical development. But there are other great problems that require them to think creatively and use numbers flexibly. Here is a small selection:

The Four 4's

Try to make every number between 0 and 20 using only four 4's and any mathematical operation (such as multiplication, division, addition, subtraction, raising to a power, or finding a square root), with all four 4's being used each time. For example

$$5 = \sqrt{4} + \sqrt{4} + \frac{4}{4}$$

How many of the numbers between 0 and 20 can be found?

Race to 20

This is a game for two people.

Rules:
1. Start at 0.
2. Player 1 adds either 1 or 2 to 0.
3. Player 2 adds either 1 or 2 to the previous number.
4. Players continue taking turns adding 1 or 2.
5. The person who gets to 20 is the winner.

See if you can come up with a winning strategy.

Painted Cubes

A 3 × 3 × 3 cube

is painted red on the outside. If it is broken up into
1 × 1 × 1-unit cubes, how many of these small cubes
have 3 sides painted? Two sides painted? One side
painted? No sides painted? What about if you start with
a larger original cube?

Beans and Bowls

How many ways are there to arrange 10 beans among 3 bowls? Try it for different numbers of beans.

Partitions

You could use Cuisenaire rods to help with this problem.
The number 3 can be broken up into positive numbers in four different ways:

1 + 1 + 1

1 + 2

2 + 1

3

Or maybe you think that 1 + 2 and 2 + 1 are the same, so there are really only three ways to break up the number.

Decide which you like better and investigate partitions for different numbers using your rules.

Websites and Books.

There are many wonderful puzzles to be found on websites and in books. Here is a small and selective selection of the best websites containing mathematical puzzles:

http://nrich.maths.org/public/
N-rich is a project run by The University of Cambridge, providing excellent puzzles and problems at different levels of difficulty for children at all school ages.

http://www.bbc.co.uk/cbeebies/numberjacks/
Numberjacks is a BBC programme for young children, introducing them to maths in a really nice way, showing the importance of patterns in children's thinking. The Numberjacks website includes a range of games and puzzles, including a version of Sudoku for children.

http://www.mathsphere.co.uk/resources/MathSphereMathsPuzzl es.htm
Another great resource for puzzles.

http://www.bbc.co.uk/schools/websites/4_11/site/numeracy.sht ml
The BBC has a range of math problems and puzzles available for different age groups: pre-school, 4-11, 11-16 and 16+

http://www.4learning.co.uk/sites/puzzlemaths/
Channel 4 also has a range of puzzles available at different levels, starting from primary Key Stage 2.

http://www.parentscentre.gov.uk
This government sponsored parent centre also contains a range of puzzles as well as information on the National Numeracy Strategy, which governs the teaching of maths in primary schools and the Framework for Secondary Mathematics.

The following is a selection of books, containing puzzles and problems. Some of the books are now a few years old, but they are all excellent and still available, from amazon uk as well as

Beans and Bowls

How many ways are there to arrange 10 beans among 3 bowls? Try it for different numbers of beans.

Partitions

You could use Cuisenaire rods to help with this problem.

The number 3 can be broken up into positive numbers in four different ways:

[ ][ ][ ]	1 + 1 + 1
[ ][ ]	1 + 2
[ ][ ]	2 + 1
[ ][ ][ ]	3

Or maybe you think that 1 + 2 and 2 + 1 are the same, so there are really only three ways to break up the number.

Decide which you like better and investigate partitions for different numbers using your rules.

Websites and Books.

There are many wonderful puzzles to be found on websites and in books. Here is a small and selective selection of the best websites containing mathematical puzzles:

http://nrich.maths.org/public/
N-rich is a project run by The University of Cambridge, providing excellent puzzles and problems at different levels of difficulty for children at all school ages.

http://www.bbc.co.uk/cbeebies/numberjacks/
Numberjacks is a BBC programme for young children, introducing them to maths in a really nice way, showing the importance of patterns in children's thinking. The Numberjacks website includes a range of games and puzzles, including a version of Sudoku for children.

http://www.mathsphere.co.uk/resources/MathSphereMathsPuzzl es.htm
Another great resource for puzzles.

http://www.bbc.co.uk/schools/websites/4_11/site/numeracy.sht ml
The BBC has a range of math problems and puzzles available for different age groups: pre-school, 4-11, 11-16 and 16+

http://www.4learning.co.uk/sites/puzzlemaths/
Channel 4 also has a range of puzzles available at different levels, starting from primary Key Stage 2.

http://www.parentscentre.gov.uk
This government sponsored parent centre also contains a range of puzzles as well as information on the National Numeracy Strategy, which governs the teaching of maths in primary schools and the Framework for Secondary Mathematics.

The following is a selection of books, containing puzzles and problems. Some of the books are now a few years old, but they are all excellent and still available, from amazon uk as well as other booksellers. The following list is ordered in increasing difficulty:

Bolt, B. (1984). *The Amazing Mathematical Amusement Arcade.* Cambridge: Cambridge University Press.

Bolt, B. (1992). *Mathematical Cavalcade.* Cambridge: Cambridge University Press.

Bolt, B. (1993). *A Mathematical Pandora's Box.* Cambridge: Cambridge University Press.

Bolt, B. (1995). *A Mathematical Jamboree.* Cambridge: Cambridge University Press.

Tanton, J. (2001). *Solve This: Math Activities for Students and Clubs.* Washington, D.C.: Mathematical Association of America.

Cornelius, M., & Parr, A. (1991). *What's Your Game?* Cambridge: Cambridge University Press.

Averbach, B., & Chein, O. (1980). *Problem Solving Through Recreational Mathematics.* Mineola, NY: Dover.

Berlekamp, E., & Rodgers, T. (1999). *The Mathemagician and Pied Puzzler: A Collection in Tribute to Martin Gardner.* Natick, Mass.: A K Peters, Ltd.

Gardner, M. (1987). *Riddles of the Sphinx: And Other Mathematical Puzzle Tales.* Washington, D.C.: Mathematical Association of America.

Gardner, M. (2000). *Mathematical Puzzle Tales.* Washington, D.C.: Mathematical Association of America.

Gardner, M. (2001). *The Colossal Book of Mathematics*. New York: W.W. Norton & Company.

Moscovich, I. (2005). *Knotty Number Problems & Other Puzzles*. New York: Sterling Publishing Co., Inc.

Berlekamp, E. R., Conway, J. H., & Guy, R. K. (2001). *Winning Ways for Your Mathematical Plays, Second Edition*. Wellesley, Mass.: A K Peters, Ltd.

Engaging in these puzzles and problems could be the greatest pathway to mathematical engagement and success that you or your child could ever embark upon. Enjoy!

9 / Making a Difference through Work with Schools

Teachers are placed are under huge pressure to prepare students for tests and many of them believe that the best way to do that is to show children lots of methods and have students practice them. This narrow version of maths is the cause of widespread low achievement in the country and, ironically, does not even result in good test performance[1]. We need to move away from the fake version of maths that is taught in classrooms and engage students in real maths, as a matter of urgency.

Parents can be extremely powerful in changing schools for the better, for a simple reason. Many schools teach a narrow and procedural version of maths because they think that is what parents want. When parents communicate to schools that they would like their children to be actively involved in maths, teachers are often very pleased to consider ways to make that happen.

In this chapter I will recommend important books and other resources as well as strategies for working with teachers and schools to make maths classrooms places where children enjoy mathematics, learn important maths, and become inspired to take it further.

Pat Kenschaft is a mathematician in the US and she identifies a "crisis" in classrooms[2], that is a very real crisis for children in the UK too. She says that parents can identify a crisis when:

- children bring home more than ten problems in a typical homework,
- children don't work on group activities in class, and
- children are not given puzzles and problems to work on.

I agree that these common characteristics of maths classrooms do contribute to the crisis that pervades across the country, and this sort of crisis can be taking place at any age level, in primary and secondary schools. In secondary schools mathematics teachers are more likely to be trained specialists than in primary schools – but they may still teach in a very narrow and damaging way. In a number of schools in England, children are taught mathematics by teachers trained in other fields, such as science and P.E. Some of these teachers are very good and I am not convinced that the only good mathematics teachers are those with higher degrees in mathematics, but some of them lack confidence and knowledge of good teaching methods and so can only repeat the flawed and narrow methods of teaching that they experienced in school. This is part of the reason for the maths crisis that is widespread and growing across the UK. The good news is that parents can be extremely powerful in bringing about change. If your child is experiencing maths lessons that are turning them away from maths, making them feel inadequate, or emphasizing rules and procedures at the expense of understanding, then it is really important to act. In the following two sections I recommend ways of working with teachers and schools, and further activities and resources that might be helpful.

1. Working with Teachers and schools.

(1) Meet with the classroom teacher.

The first and most important action to take in order to improve your child's maths teaching at school is to contact their teacher. All teachers want children to do as well as possible and even if they are using out-dated and unsuccessful methods then they will be doing so with the best of intentions. So my first recommendation is to get to know your child's maths teacher and discuss your child's needs with them. As an experienced teacher myself I know that all teachers are very concerned to please parents and if a parent contacts them to discuss teaching they will probably be very anxious to make sure that the meeting goes well. My advice is always to approach teachers with sensitivity and friendliness, if we are to improve the educational opportunities of our children then we need to work *with* teachers, not against them.

In the following section I have listed some questions and comments that you could discuss with your child's classroom teacher. Your conversation will probably be different depending on whether the teacher is a primary school teacher, who teaches all subjects, or a secondary teacher who is trained in mathematics. We know that many primary school teachers may have found maths difficult themselves and probably feel much more comfortable with other subject areas. Primary school teachers may also feel fully stretched in preparing all of the different subject lessons that they teach, and may not find the time to make maths as interesting as other subjects. Secondary teachers are specialists and it is fair to assume that they enjoy maths. Their comfort and confidence with maths may change the way they think about the issues you want to discuss with them. With any teacher an open discussion about issues that are concerning you, even when there is disagreement, is by far the best approach. Here is a list of questions that might be helpful to use:

- Emily (substitute your child's name) has been having problems with maths – she doesn't enjoy maths in school, although she does enjoy doing mathematical work at home. Can I ask you what maths approach you use in school? As that might help me work out ways to help her.

The teacher will probably tell you that they follow the directives set out by the government. In primary schools this is the National Numeracy Strategy (http://www.standards.dfes.gov.uk/primary/) and in secondary schools the Framework for Secondary Mathematics (http://www.standards.dfes.gov.uk/secondary/framework/maths/fwsm). The secondary framework has recently been rewritten to include a greater emphasis on problem solving in maths. If your child's school is not emphasizing problem solving in maths, or students are not working in groups together, whether at primary or secondary levels, it is likely that they are not teaching in a way that is consistent with the government's framework.

However your child's teacher responds I would follow up with your main concern, for example:

- Emily really enjoys a thinking approach to maths, where she is able to discuss mathematical ideas, and work on maths puzzles and problems, is it possible that class could involve more puzzles and problems, and more time for discussion, with less repetition?

This is the critical question that could go in all sorts of directions. The teacher may be open to such a suggestion or she may argue against it.

If the teacher argues against a problem-solving approach then one strategy is to ask the teacher whether she or he has read about the importance of such approaches and if there is any space to work in this way in class. Another is to ask whether the teacher would be open to reading a book with you, that you could then discuss together. In the next section I recommend a number of books for different age levels that you could discuss with the teacher, this book is also one the teacher may be willing to read. All of the books I list are ones that could open a teacher's eyes to more effective ways to teach mathematics, that focus on understanding.

It may be that the teacher does not support the approach used by the school but feels compelled to use it because it was chosen by others at the school. In a primary school it would then be appropriate to talk to the maths co-ordinator, about the approach used in maths, and possibly include the head-teacher in such a meeting. In secondary schools the person to speak to next is the mathematics head of the department and pose the questions listed above.

It may be that the teacher is open to using a different mathematics approach, with more problem-solving and discussion but does not know how to, or does not feel secure with such an approach. I would be very supportive and ask if the school is encouraging the teacher's development, by, for example, providing mathematics professional development opportunities. Additionally you could recommend some books to read, and websites to visit – in the resources section I recommend some books that incorporate a number of activities that teachers could use in class. Patricia Kenschaft tells the story of a mathematics educator needing to visit her son's 5th grade teacher as he was bringing home 30 exercises a night for homework; when the mother arrived at the meeting, ready for a confrontation, the teacher said "I'm so glad to see you" as she had been nervous about the maths she was teaching and she was looking forward to talking about it with someone. They ended up planning lessons together and the teacher changed her approach to be much more effective[3]. You may not feel you have the expertise to help the teacher in the way that a mathematics educator did but the teacher may welcome a friendly person to talk through ideas with. The teacher may also request that you talk to the department chair or head-teacher, as parents can be influential in getting extra support for teachers.

Whatever the reason the teacher gives, for using an approach that you are unhappy with, it is important to get this information as it will help you know who to talk to next. It is likely that the teacher's replies will lead you to a conversation with the head of department, the mathematics coordinator or the head-teacher, as I discuss shortly.

- What do you think are the best ways to engage students?

This is a good question for both primary and secondary teachers. Primary teachers have a great deal of expertise in teaching young children and usually have a wealth of knowledge in ways to organize and present new ideas so that children are engaged and stimulated, but the good ideas they use in teaching literacy and arts are often put aside when it is time to teach maths. If a teacher is skilled in teaching other subject areas s/he can probably be encouraged to apply similar methods to maths and the books listed in the resources section will help her do that. It can also be a good question to ask secondary teachers. You may find that the ideal approach for many secondary teachers is one that is thinking-based but that they feel they cannot use such an approach because of certain constraints. For example you may find that they would give students more opportunities to explore mathematics, think and discuss ideas, if they did not have narrow tests to prepare for if they felt children – or parents – would be open to such an approach. A number of teachers at the secondary and primary level only teach students to rehearse standard methods and do not give students space or time, to use or apply them, or to discuss their ideas, because they feel pressured by tests and do not think there is time to do anything else. These teachers may not be aware that studies of teachers who spend more time on exploration and mathematical use and less time on practice, have found that students score as well, or better on even the narrowest of tests and significantly better on more probing assessments[4]. Students may 'cover' less but if they understand their work then they are likely to do well on tests, even when the tests differ in nature from the work they may be used to.

- Could I visit the class and sit in on a lesson?

Visiting maths class can be tremendously helpful as it gives you an opportunity to see the difficulties your child is facing, and it provides the best opportunity for discussions with the teacher. If the teacher is willing to have you visit class then it is probably best to do this before having discussions about their approach. When discussing the visit ask the teacher what sort of lesson would be best to visit, and what role you should play. Some teachers might like you to join in and sit with children, discussing ideas with them and giving them help when you can, others would prefer that you sit apart from the children and simply watch.

- What methods do you use to promote 'assessment for learning' in your class?
- Could I receive a copy of the learning goals that students are given, to help at home?

Your child's teacher will probably know about 'assessment for learning' as it is part of national policy for all schools in England and Wales, but they may not be using the approach as set out in chapter 4. I recommend talking to the teacher about the incredible results that have been achieved when children work towards clear learning goals that they know about in advance. Also ask the teacher if s/he could share the criteria with you so that you may help your child at home.

The government has just released a DVD showing good strategies to use in mathematics, in primary schools (although many would be effective in secondary schools too). This is called *'Developing assessment for learning in mathematics – classroom practice in action'*, with the reference number: 00101-2008BKT-EN. (http://www.standards.dfes.gov.uk/primary/publications/mathematics/developing_afl/)

In secondary schools much of the responsibility for decisions about the maths teaching approach falls to the head of department and I would advise speaking to the head of department either after or as an alternative to your child's maths teacher. The questions and discussion would be similar, but the head of department provides both you and the teacher with an alternative person, or an extra person, to talk to. In primary schools the equivalent person is the maths coordinator.

(2) Meet with the head-teacher.

In some situations it may be appropriate to meet with the head-teacher, or the deputy or assistant head-teacher who has responsibility for curriculum. Such a meeting would be appropriate if your child's class teacher would like some support from their leaders, or if the teacher will not consider non-traditional approaches. When maths teachers in secondary schools do not engage students in problem-solving and working in groups, head-teachers are often aware of this and concerned, as it means that maths teaching is out of synch with the rest of the teaching in the school. They would then be very pleased to hear from you, as it helps them work with the teachers concerned.

The head or deputy head teacher is also a good person to consider the school's assessment approach and decisions about ability grouping. In the following I suggest discussions that you may have with the head-teachers around the three critical issues of teaching approach, assessment and grouping.

- The Teaching Approach.

If meeting with the head teacher about the maths teaching approach I would ask similar fact-finding questions that I recommended for the teacher – who decided upon the school's approach? Have other approaches been considered that involve students more actively? I would also find out about school performance in mathematics, compared with other schools and compared with other subjects at the school. The head teacher will have a lot of data that s/he can show you. Ask the head teacher not only about test results but the nature of the tests used and whether opportunities are taken up for the use of broad assessments. Ask if teachers have regular professional development opportunities that help them learn how to teach mathematics well.

- Assessment.

It is critical that schools use good assessments to test students, that are different from SATs, which are insufficiently broad. A head-teacher should know whether the school is or could use good assessments. The National Foundation for Educational Research (NFER) has recently developed a range of on-line assessments that children in primary schools take in their classrooms[5]. These give schools helpful diagnostic information on students' strengths and weaknesses – showing what students did wrong and why. The tests are available for schools use but head-teachers may be unaware of their existence. Broad assessments that assess problem solving seem harder to find in the UK. In the US the

Balanced Assessment Group at Harvard University, developed around 300 high quality assessments that are freely available on-line[6], and highly applicable to classrooms in the UK. Again these could be brought to the attention of the head-teacher.

The head-teacher will know about the 'assessment for learning' approach and how the school is using it. If you want to help your own child by discussing criteria that they should be aiming for in maths, then you could ask the head-teacher how you could get copies of criteria for each unit of work, as such learning goals are a critical part of the approach.

- Grouping.

It is wise to ask the head-teacher about the use of ability grouping in the school. Are students sorted into different classes now, or will they be in the near future? Does the head-teacher know that research shows that mixed ability classes tend to result in higher achievement? Has the head teacher read recent reviews of research on grouping structures, at primary and secondary levels, given in the resources section below? Would the head-teacher be willing to read chapter 5 of this book?

Finally, you could ask the head-teacher what professional development the school is providing teachers to help them know about ways to teach and assess well.

It may seem daunting to arrange a meeting with your child's teacher or the head-teacher, but don't be put off, your intervention might be critical in shaping your child's mathematical future and your interest and enthusiasm will probably be welcomed by the school.

(3) Attend PTA meetings.

The parent teacher association is an excellent place to take concerns about mathematics teaching in a school. Instead of arriving at such organizations with a list of complaints, as some people do, I would recommend asking the PTA to consider a list of questions. This list is one that I like that was put together by Ruth Parker:

1. Is our school's mathematics approach teaching children to think and reason and make sense of the mathematics they are learning?
2. Is practice with skills provided in engaging, challenging and mathematically important contexts?
3. Is persistence valued over speed?
4. Are problem solving and a search for patterns at the core of all that children are asked to do?
5. Is numerical reasoning emphasized?
6. Does the mathematics approach emphasize that there is almost always more than one way to solve a mathematics problem?
7. Does it present mathematics as relationships to be understood rather than recipes to be memorized?
8. Are children the ones who are doing the thinking and sense making?

The PTA should welcome a discussion of such issues and could draw up some recommendations for the school. If you would like to have more influence you could run for election to the PTA.

Parents can also play a big role in helping the PTA raise funds for professional development opportunities and other critical resources that will help the maths teaching.

(4) Speak to a member of the Governing Body.

In England schools are over-seen by the governing body. Governors of state schools are responsible for raising school standards through their three key roles of setting strategic directions, ensuring accountability, and monitoring and evaluating school performance. In most schools at least one-third of the governors are parent governors, others are staff governors, including the head who is usually a member (unless he or she declines the opportunity) and members of the community. If you feel that discussions with teachers or head-teachers are not progressing, or not resulting in improvements, then it would be reasonable to ask to meet with a parent governor and set out your concerns about maths teaching.

(5) *Provide other mathematical opportunities.*

This suggestion may sound strange if you don't feel mathematically confident yourself but even if you do not I suggest starting a lunchtime or an after school maths club. You could provide mathematical settings and mathematical puzzles and games, such as those described in the previous chapter, the school should be able to provide funds for these, or the PTA could raise them. Children will enjoy working on maths with others and you don't need to be an expert on maths to set it up for them. You may be able to involve other parents or teachers in the club. This could also be a place that children get help with their homework or work on their homework with others. Most importantly it would be a place where children could see mathematics in its authentic form and where they could enjoy and be supported in mathematical work. The event could be advertised not as being for high flyers, but for anyone who wants to enjoy maths, and to talk with others about maths, even if they have never enjoyed it before.

2. Books and Other Resources.

Websites

www.mec-maths.org The Mathematics Education Collaborative (MEC) is a non-profit organization that works with parents and the public in support of quality mathematics in schools. MEC's *Supporting School Mathematics* series consists of Presenter's Guides for six interactive maths workshops for parents and teachers. MEC's sessions are designed to help both parents and teachers understand important issues in mathematics education, so although they are in the US they could be very helpful to families in other countries. The workshops provide many mathematical games and activities that families can play and do together. As a parent you can ask the head-teacher or PTA to make these resources available so that you, interested teachers, and/or local education authority leaders can offer the workshops locally.

www.mathematicallysane.com. This is the website that was set up to counter the movement in the US that campaigns against non-traditional forms of mathematics teaching. Because anti-reformers are very active in telling lies to stop anybody moving forward from traditional approaches, this web-site was set up to include real data and discussions on good mathematics approaches.

http://www.parentscentre.gov.uk This government sponsored parent centre contains advice on ways to help children, and information on the teaching in schools, including information on the National Numeracy Strategy and the Secondary Mathematics Strategy, which governs the teaching of maths in primary and secondary schools.

Other websites providing mathematics puzzles and problems are given in chapter 8.

Books & Articles.

Learning Mathematics.

- Sue Atkinson (ed). (1992). *Mathematics with Reason: The Emergent Approach to Primary Mathematics*. Hodder Arnold H & S.

This book explains how primary teaching of mathematics can be brought in line with research on how children think and learn best. The book is now a few years old but it is still highly relevant and is still available on amazon UK.

- Derek W. Haylock (2005). *Mathematics Explained for Primary Teachers*.

A clear explanation of all of the mathematics in primary schools, a great resource for teachers or parents of primary age children. It is published by Sage Publications.

- Kathy Richardson's 'Developing Math Concepts' series, is written by a leading educator from the US and provides a great resource for primary teachers and parents of primary age children, particularly students from reception to year 4, showing an excellent approach to the teaching of maths. These books may need to be bought from book-sellers in the US, (available via amazon UK).

Richardson, K. (1998). *Developing Number Concepts: Counting, Comparing, and Pattern*. Dale Seymour Publications
Richardson, K. (1998). *Developing Number Concepts: Addition and Subtraction*. Dale Seymour Publications
Richardson, K. (1998). *Developing Number Concepts: Place Value, Multiplication and Division*. Dale Seymour Publications
Richardson, K. (1999). *Understanding Geometry*. Dale Seymour Publications

- Ball, D. L. (1993). With an Eye on the Mathematical Horizon: Dilemmas of Teaching Elementary

Mathematics. *The Elementary School Journal,* 93(4), 373-397.

This is a journal article that is intended for primary mathematics teachers and researchers. It includes discussion of the important decisions an excellent elementary teacher in the US, Deborah Loewenberg Ball, (now university dean) made when teaching mathematics and is well worth reading, especially by teachers who will understand the dilemmas the author refers to. The article is available on the author's own website (http://www-personal.umich.edu/~dball/articles/index.html)

- Cathy Fosnot and Maarten Dolk's 'Young Mathematicians At Work' (Heinemann) series. A series of 3 books in which the authors focus on the ways children between 4 and 8 develop a solid understanding of maths.

Fosnot, C. & Dolk, M. (2001) *Young Mathematicians at Work: Constructing Number Sense, Addition, and Subtraction.* Heinemann Publications

Fosnot, C. & Dolk, M. (2001) *Young Mathematicians at Work: Constructing Multiplication and Division.* Heinemann Publications

Fosnot, C. & Dolk, M. (2002) *Young Mathematicians at Work: Constructing Fractions, Decimals and Percents.* Heinemann Publications.

- John Van de Walle and Lou Ann Lovin's series giving practical advice and guidance for teaching primary age students.

Van de Walle, J. and Lovin, L. (2006). *Teaching student centered mathematics*, grades K-3. Boston, MA: Pearson

Van de Walle, J. and Lovin, L. (2006). *Teaching student centered mathematics*, grades 5-8. Boston, MA: Pearson

• Hiebert, J., Carpenter, T., Fennema, E., Fuson, K., Wearne, D., Murray, H., et al. (1997). *Making Sense: teaching and learning mathematics with understanding*. Portsmouth, NH: Heinemann.

This award winning book gives teachers – and parents – ways to promote and defend classrooms that focus on mathematical understanding. Four separate research programs are drawn upon as the authors describe the essential features of effective maths classrooms.

• Hiebert, J., Carpenter, T., Fennema, E., Fuson, K., Human, P., Murray, H., et al. (1996). *Problem Solving as a Basis for Reform in curriculum and Instruction: The Case of Mathematics*. Educational Researcher, 25(4), 12-21.

A journal article by the same authors that shows the importance of a problem solving approach to mathematics.

• Ma, L. (1999). *Knowing and Teaching Mathematics: teaching Understanding of Fundamental Mathematics in China and the United States*: Lawrence Erlbaum Association.

In this book Liping Ma compares the understanding of maths shown by two groups: American and Chinese elementary teachers and shows the importance of understanding mathematical concepts. Ma also contrasts different ways to teach primary maths, with and without conceptual understanding.

- Kilpatrick, J., Swafford, J., & Findell, B. (Eds.). (2001). *Adding it up: Helping children learn mathematics.* Washington, DC: National Academy Press.

This book presents the findings of a special mathematics committee formed by the National Research Council in the US and gives advice on the ways teaching should change to offer better mathematical experiences for primary age children.

- Stigler, J., & Hiebert, J. (1999). *The Teaching Gap: Best Ideas from the World's Teachers for Improving Education in the Classroom.* New York: Free Press.

This readable book offers an analysis of the best teaching methods from around the world, with detailed accounts of the video evidence from the Third International Mathematics and Science Study (TIMSS).

Assessment for Learning.

- Black, P., Harrison, C., Lee, C., Marshall, B., & Wiliam, D. (2003). *Assessment for Learning*: Open University Press.

This book summarizes the research showing the effectiveness of the 'assessment for learning' approach and follows teachers who were using the approach, giving helpful advice for teachers and teacher leaders who want to implement the approach.

Ability Grouping

- Blatchford, P., Hallam, S., Ireson, J., Kutnick, P. & Creech, A. (2008). Research Survey 9/2 *Classes, Groups and Transitions: Structures for Teaching and Learning Reports.* Commissioned as evidence to The Primary Review. The University of Cambridge. (available from http://www.primaryreview.org.uk/Publications/Interimr eports.html)

This is a recent and important review on grouping strategies in mathematics, in primary classrooms, commissioned for The Primary Review.

- Kutnick, P., Sebba, J., Blatchford, P., Galton, M. and Thorp, J. (2005) *The Effects of Pupil Grouping: Literature Review.* London, DfES. Research Report 688.

A report, commissioned for the (then) Department for Education and Science (DFES), summarizing research on the impact of ability grouping.

- Dunne, M., Humphreys, S., Sebba, J., Dyson, A., Gallannaugh, F. & Muijs, D. (2007) *Effective Teaching and Learning for Pupils in Low Attaining Groups* Nottingham: DfES Publications.

A report, commissioned for the (then) Department for Education and Science (DFES), revealing the high number of students placed into the wrong ability groups.

- Boaler, J., Wiliam, D., & Brown, M. (2000). Students' experiences of ability grouping – disaffection, polarization and the construction of failure. *British Educational Research Journal*, 26, 5, 631-648.

Research paper, recommended on the Government's Strategies site, on the impact of ability grouping in secondary mathematics classrooms.

In Conclusion.

Maths is a painful experience for far too many children and adults. Instead of introducing children to the science of patterns, a fascinating set of ideas, and a way of interpreting the world, children are led through hours of repetitive procedures that they come to believe is maths. This misrepresentation of the subject, is nothing short of scandalous and it is something we must all work to fight. For mathematics is becoming more and more important for the future of society and our children. Without mathematical know-how young adults are disadvantaged and vulnerable.

This book is all about helping children, and adults, have a positive relationship with maths, feel capable in the face of mathematical problems, and ultimately contribute to a society that can take scientific, medical and technological work forward. I have talked in this book about the importance of raising mathematical achievement and interest for the future of *society*, but perhaps my greatest motivation in writing this book comes from my desire to improve children's experiences in classrooms – to take the fear and boredom out of the subject, and replace it with excitement and interest. Mathematicians will tell you that the subject they care so much about is a living, connected and *beautiful* subject. This book is about giving all children, not only an elite few, the same important insights. Whether you are a parent or an educator, you can be extremely powerful in giving children a different view of mathematics and a better relationship with the subject. As you do this you will not only be giving them a brighter future but an amazing set of tools and insights with which to interpret, make sense of and enjoy the world around them.

Appendix

Solutions to the Mathematics Problems

Introduction

The Skateboard Problem

A skateboarder holds onto the merry-go-round pictured below. The platform of the merry-go-round has a 7-foot radius and makes a complete turn every 6 seconds. The skateboarder lets go at the 2 o'clock position in the picture at which time she is 30 feet from the padded wall. How long will it take the skateboarder to hit the wall?

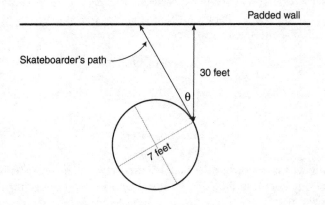

A Solution:

The first step is to find out how far the skateboarder travels after letting go. In other words, we need to figure out the distance AB in the drawing below.

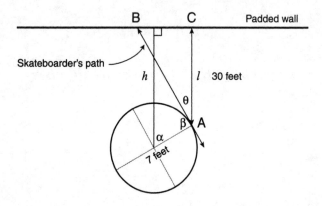

To do this we first need to work out what the angle Θ is so we can use the properties of right triangles. To do this, draw the line h that passes through the center of the merry-go-round and meets the padded wall at a right angle. Since the skateboarder is at the "2-o'clock" position, which is 1/6 of the way around a clock, the angle α is 1/6 of 360°. So $\alpha = 60°$. The angle $\beta = 90°$, since a tangent line of a circle always meets the radius at a right angle. Finally, $\alpha + \beta + \Theta = 180$, since they are same-side interior angles between the parallel lines h and l. Therefore $\Theta = 30°$. This means that triangle ABC is a 30-60-90 triangle! Using the side relations of 30-60-90 triangles, we find:

$$BC = \frac{30°}{\sqrt{3}} = 10\sqrt{3} \text{ feet}$$

and

$$AB = 20\sqrt{3} \text{ feet}$$

So now we know how far the skateboarder travels. The next step is to figure out how fast she is travelling. The merry-go-round makes a complete turn every 6 seconds. In a complete turn, the skateboarder travels the entirety of the circumference, which is

$$C = 2\pi \,(7) = 43.98 \text{ ft.}$$

So the skateboarder is travelling at $43.98/6 = 7.330$ feet per second.

Since

$$\text{distance)} = \text{(rate) (time)}$$

then

$$\text{(time)} = \text{(distance)/(rate)}$$

So the time it takes the skateboarder to reach the wall is

$$\frac{20\sqrt{3}}{7.330} = 4.726 \text{ seconds}$$

The Chessboard Problem

Solution:
What makes this problem difficult is all the different sizes of squares on a chessboard, from the smallest 1×1 squares, to overlapping 2×2 squares, all the way up to the entire chessboard, which is itself an 8×8 square.

In situations like this, it is often helpful to be organized. One way to organize the problem is to count all the different sizes of squares separately. So, let's start with the 1×1 squares. There are 8 rows and 8 columns on the board, so there are 64 of these. Next, let's look for the 2×2 squares. These are more difficult, as they can overlap, as the two grey squares below do:

Even overlapping squares have distinct centre points, though, so an easy way to keep track of the overlapping squares is to mark the centre of each square with a dot. Here are some examples of squares with their centre points marked:

Marking all the centre points of the 2 × 2 squares, we get a grid of centre points:

Notice this is a 7 × 7 grid of points. So there are 49 2 × 2 squares.

To keep track of the 3 × 3 squares, we can also mark the centre points, as in the following few examples:

When we draw all of the centre points for the 3 × 3 squares, we get a picture that looks like this:

This is a 6 × 6 grid of points, so there are 36 3 × 3 squares. Continue this process until you have counted the 4 × 4 squares, the 5 × 5 squares, and so on. The total number of squares is then

$$8^2 + 7^2 + 6^2 + 5^2 + 4^2 + 3^2 + 2^2 + 1 = 204$$

This process of counting works for any size of chessboard In general, for an *n*-by-*n* chessboard, the number of 1 × 1 squares is n^2. The number of 2 × 2 squares is $(n - 1)^2$ and so on. So the total number of squares is

$$n^2 + (n - 1)^2 + (n - 2)^2 + \ldots + 3^2 + 2^2 + 1$$

Chapter 3

The Railside Pattern Problem

Juan's problem: "See if you can work out how the pattern is growing and the algebraic expression that represents it!"

Solution:

A good way to solve this problem is to look at how each part is growing separately. One way to see it is this: in the diagrams above it looks like the white squares on the left are growing by one each time the pattern proceeds to the next step as are the white squares on the right. The solid black square on the right and the solid black square on the bottom don't seem to change. And the rectangle of gray squares is growing by one in each of its dimensions. To come up with a formula we need to number each picture. Let's call the first picture "$n = 1$" and the next picture "$n = 2$." So now let's try to say how many of each square there are in terms of n. The white squares on the left and the white squares on the right always have one more than n, so these can each be represented by the expression $(n + 1)$. The black squares on the bottom and on the right are always just 1 no matter what n is, so these can be represented by the expression 1. Finally, the rectangle of gray squares has a width of n and a height that is two more than n, or $(n + 2)$. So the number of squares in this rectangle can be represented by the width times the height or the expression $n(n + 2)$. So the total number of squares on the nth day is

$$(n + 1) + (n + 1) + 1 + 1 + n(n + 2)$$
$$= n + 1 + n + 1 + 1 + 1 + n^2 + 2n$$
$$= n^2 + 4n + 4$$

Interesting note: this expression factors as the perfect square $(n + 2)^2$, which means that every arrangement of squares can be arranged as a square. Try to see how they rearrange. This could lead to a different way of solving the problem.

The Amber Hill Question

Helen rides a bike for 1 hour at 30km/hour and 2 hours at 15km/hour. What is Helen's average speed for the journey?

Solution:

"Average speed" is one of those tricky expressions in word problems, because it can be interpreted to mean different things. The most natural interpretation is "If you were traveling at a *constant* speed, how fast would you be going to cover the same distance in the same amount of time?" This interpretation allows you to work out the total distance traveled and the total time spent. Then the average speed is (total distance) / (total time). In this problem. Helen travels 30 km for the first hour and 15 × 2 = 30 km for the second and third hours. So the total distance is 30 km + 30 km = 60 km. The total time is 1 hour + 2 hours = 3 hours. So the average speed is (total distance) / (total time) = (60 km) / (3 hours) = 20 km/hour.

Chapter 7

The Staircase Problem

In this task students were asked to determine the total number of blocks in a staircase that grew incrementally from 1 block high, to 2 blocks high, to 3 blocks high, and so on, as a move toward predicting a 10-block-high staircase and a 100-block high staircase, and, finally, algebraically expressing the total blocks in any staircase. Students were provided with a box full of linking cubes to build the staircases if they wished.

A 4-block-tall staircase: total blocks = 4 + 3 + 2 + I = 10

Solution:
There are many ways to "see" the growth in such a staircase. One of the most elegant is to think about two copies of the staircase fitted together as shown at the top of the next page:

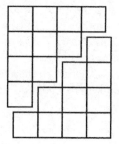

These fit together perfectly into a 4×5 rectangle. Since we put two staircases together, the total number of squares in our original staircase is $(4 \times 5)/2 = 10$. In general, two of the nth staircases can he put together to make an $n(n + 1)$ rectangle. Again, since two staircases were put together, the total number of squares in one is $n(n + 1)/2$. This is sometimes called the nth triangular number, because a staircase is triangular in shape. Using this formula we can work out how many squares are in each staircase. For example, the 10th staircase ($n = 10$) has $10(10 + 1)/2 = 55$ squares. The 100th staircase has $100(100 + 1)/2 = 5050$ squares. Rumor has it the famous mathematician Karl Gauss worked out in this way when his teacher asked the class to add up the numbers 1 through 100. Can you work out why this would give you the correct answer for that sum?

Alonzo's Staircase Problem

5b   5 + 9 = 14 *blocks*   5 + 9 + 13 = 27 *blocks*

Solution:
Alonzo's problem is like the staircase problem, except the staircase comes out in four directions from a center point. Again, we have numerous ways to count how many squares there are in this shape.

One method is to use the solution to the staircase problem. Each of Alonzo's staircases is formed by 4 copies of the staircases from the above problem, plus the squares in the center column, where the 4 copies are attached. So the number of squares is $4n[(n + 1)/2] + n$. The first term represents the four staircases, and the n is the column of squares in the middle. This last term is n because Alonzo's shape is 1 high in the first case, 2 high in the second case, and in general n high in the nth case. This formula simplifies as follows:

$$4n[(n + 1)/2] + n$$
$$= 2n(n + 1) + n$$
$$= 2n^2 + 2n + n + n$$
$$= 2n2 + 3n$$

Whenever you come up with a general algebraic formula, it's good to make sure it works for the small cases that you understand. Can you try this formula to see if it works for the initial cases of Alonzo's staircase pictured above? Can you come up with this formula by grouping the cubes a different way?

Cowpens and Bullpens Problem

During an activity called "Cowpens & Bullpens" students had to determine how many lengths of fencing were required to contain an increasing number of cows, given certain fencing parameters.

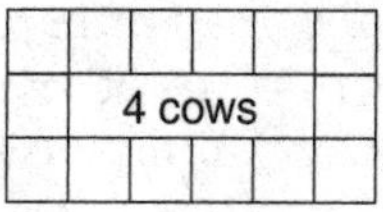

4 cows

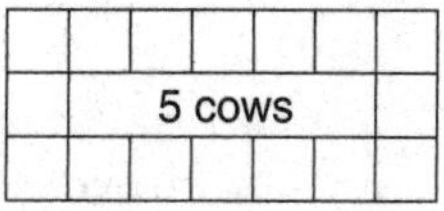

5 cows

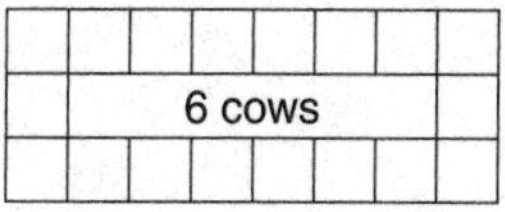

6 cows

Solution:

There are many ways to see this problem's pattern, as with others in the book. One way is to count the number of sides of fencing on each side of the cows, then add the corners. The first interesting thing to notice is that the number of sides of fencing above and below the cows is the same as the number of cows, and the number of sides of fencing to the left and right is always 1. So for the first picture above there are $4 + 4 + 1 + 1 + 4 = 14$ sides of fencing, where the first two 4's come from the fencing above and below the cows, the 1's come from the fencing to the left and the right of the cows, and the last 4 comes from the 4 corners. For the next picture there are $5 + 5 + 1 + 1 + 4$ for similar reasons. In general, there are $n + n + 1 + 1 + 4$ sides of fencing, or $2n + 6$ sides.

Chapter 8
Problems from Sarah Flannery's Book

The Two-Jars Puzzle

Given a 5-litre jar and a 3-litre jar and an unlimited supply of water, how do you measure out 4 litres exactly?

Solution:

There are many ways to solve this problem—actually an infinite number! Here's one way. Fill the 5-litre jar. Pour the water from the 5-litre jar into the 3-litre jar until the 3-litre jar is full. Now you have 2 litres of water remaining in the 5-litre jar. Dump out the water in the 3-litre jar. Then, put the 2 litres of water from the 5-litre jar into the 3-litre jar. Now fill up the 5-litre jar completely. Pour water from the 5-litre jar into the 3-litre jar until it is full. Since there were already 2 litres of water in the 3-litre jar, you have poured exactly 1 litre out of the 5-litre jar. So, there are exactly 4 litres remaining in the 5-litre jar.

If that was confusing, here's a table showing how many litres of water are in each jar at each step, with explanations of what happened at each step:

3-litre jar	5-litre jar	What happened:
empty	empty	Nothing yet!
empty	5 litres	Filled the 5-litre jar to the brim.
3 litres	2 litres	Filled the 3-litre jar with the 5-litre jar.
empty	2 litres	Dumped out the 3-litre jar.
2 litres	empty	Poured the contents of the 5-litre jar into the 3-litre jar.
2 litres	5 litres	Filled the 5-litre jar to the brim.
3-litres	**4 litres!!**	Topped off the 3-litre jar using contents of the 5-litre jar.

An interesting follow-up question: is this the least number of steps, or is there a faster way to measure 4 litres? Another follow-up question: are there any quantities you can't make with these two jars?

The Rabbit Puzzle

A rabbit falls into a dry well 30 metres deep. Since being at the bottom of a well was not her original plan, she decides to climb out. When she attempts to do so, she finds that after going up 3 metres (and this is the sad part), she slips back 2. Frustrated, she stops where she is for that day and resumes her efforts the following morning—with the same result. How many days does it take her to get out of the well?

Solution:
This is a classic example of a "trick" problem. Even if you see the trick, it is easy to make a mistake. One good thing to notice is that the act of climbing and sliding can be greatly simplified. Instead of thinking of it as going up 3 metres and down 2 metres every day, you can just think about it as going up 1 metre. So every day the rabbit goes up 1 metre—which should mean it takes her 30 days to get out, 1 metre each day. But the reason it does not is that on the last day the rabbit actually gets out, she doesn't slip back down 2 metres. So the rabbit saves itself 2 days, and it only takes 28 days. Can you work out other ways of saying this and seeing why it is actually 28 days?

The Buddhist Monk Puzzle

One morning, exactly at sunrise, a Buddhist monk leaves his temple and begins to climb a tall mountain. The narrow path, no more than a foot or two wide, spiralled around the mountain to a glittering temple at the summit. The monk ascended the path at varying rates of speed, stopping many times along the way to rest and eat the dried fruit he carried with him. He reached the temple shortly before sunset. After several days of fasting he begins his journey back along the same path, starting at sunrise and again walking at variable speeds with many pauses along the way, finally arriving at the lower temple just before sunset. Prove that there is a spot along the path that the monk will occupy on both trips at precisely the same time of day.

Solution:
This problem is part of a beautiful class of problems called "fixed-point theorems." If you like this one, there are many others like it! One of the prettiest ways to see this solution is to imagine the graph of the monk's journey, with time on the x-axis, and position on the y-axis. So his journey on the first day might look something like this:

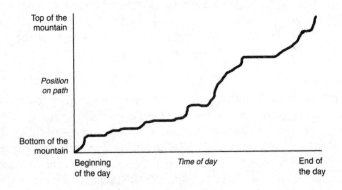

Then, on the same graph we can represent his journey down the mountain:

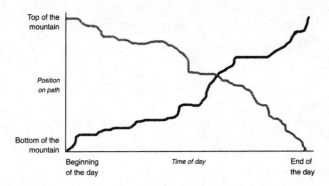

Note, these two paths may look very different, because he might choose to speed up or slow down at different times. However, since the first path must go from the lower left corner to the upper right corner, and the second path must go from the upper left corner to the lower right corner, they must cross somewhere. And, as you can see, they do cross somewhere. This point marks the time of day and the location in which the monk was in the same place at the same time on both days.

The Chessboard Problem

Solution:
(See Introduction solutions.)

The Four 4's

Try to make every number between 0 and 20 using only four 4's and any mathematical operation (such as multiplication, division, addition, subtraction, raising to a power, or finding a square root), with all four 4's being used each time. For example

$$5 = \sqrt{4} + \sqrt{4} + \frac{4}{4}$$

How many of the numbers between 0 and 20 can be found?

Solution:
There are many ways to make some numbers with four 4, but for some other numbers it is much more difficult. Here's one set of solutions for the numbers 1 through 20:

$$0 = 4 - 4 + 4 - 4$$
$$1 = 4/4 + 4 - 4$$
$$2 = 4/4 + 4/4$$
$$3 = \sqrt{4 \times 4} - 4/4$$
$$4 = \sqrt{4} + \sqrt{4} + 4 - 4$$
$$5 = \sqrt{4} + \sqrt{4} + 4/4$$
$$6 = 4 + \sqrt{4} + 4 - 4$$
$$7 = 4 + \sqrt{4} + 4/4$$
$$8 = 4\sqrt{4} + 4 - 4$$
$$9 = 4 + 4 + 4/4$$
$$10 = 4 \times 4 - 4 - \sqrt{4}$$
$$11 = \frac{\sqrt{4}(4! - \sqrt{4})}{4}$$
$$12 = 4(4 - 4/4)$$
$$13 = \frac{\sqrt{4}(4! + \sqrt{4})}{4}$$
$$14 = 4! - 4 - 4 - \sqrt{4}$$
$$15 = 4 \times 4 - 4/4$$
$$16 = 4 \times 4 + 4 - 4$$
$$17 = 4 \times 4 + 4/4$$
$$18 = 4! - \sqrt{4} - \sqrt{4} - \sqrt{4}$$
$$19 = 4! - 4 - 4/4$$
$$20 = 4 \times (4 + 4/4)$$

Notice for some solutions, I used the factorial notation, the exclamation point (!), after some numbers. This operation multiplies the number by every positive integer less than it. So 4! = 4 × 3 × 2 × 1 = 24. For the numbers where I used a factorial operation, do you think it's possible to find solutions without it?

Race to 20

This is a game for two people.

Rules:
1. Start at 0.
2. Player 1 adds either 1 or 2 to 0.
3. Player 2 adds either 1 or 2 to the previous number.
4. Players continue taking turns adding 1 or 2.
5. The person who gets to 20 is the winner.

See if you can come up with a winning strategy.

Solution:
One of the things you may notice after playing this game for a while is that if you can get to 17, you are the winner. This is because no matter what your opponent adds, whether it be 1 or 2, on your next turn you will be able to get to 20. So getting to 17 is just as good as getting to 20. This idea can extend to even lower numbers. Here, 17 is a good number for you to get to because it is 3 away from 20, just one more than your opponent can add. Keeping 3 away from these "good numbers" is the trick. By similar reasoning, getting 14 makes you the winner, because no matter what your opponent adds, on your next turn you will be able to get to 17, and then you already know what you need to do to get to 20. Similar reasoning applies to 11, 8, 5, and so on, down by 3s. So now start at the beginning: can you come up with a winning strategy if you go first? What about if you go second? If you like, make up some variations of this game that work differently, then try to find the strategy.

Painted Cubes

A 3 × 3 × 3 cube

is painted red on the outside. If it is broken up into
1 × 1 × 1-unit cubes, how many of these small cubes
have 3 sides painted? Two sides painted? One side
painted? No sides painted? What about if you start with
a larger original cube?

Solution:

The 3 × 3 cube has 1 cube right in the middle, with 0 sides painted.
It has 6 cubes with 1 side painted, 1 in the center of each of the six
faces. It has 12 cubes with 2 sides painted, 1 in the center of each edge
of the large cube. And it has 8 cubes with 3 sides painted, 1 in each
of the 8 corners of the large cube.

In general, for an $n \times n \times n$ cube, let s think about how to count all
the cubes with 0 sides painted. Imagine removing the entire outer
layer of small cubes. You'll be left with a cube in the center, but each
of its dimensions will be shrunk by 2, because a layer of cubes has been
removed on both sides. So it's an $n - 2$ by $n - 2$ by $n - 2$ cube. So it has
$(n - 2)^3$ little cubes. For the cubes with 1 side painted, these are on the
interior of each face. By similar reasoning as above, this is an $n - 2$ by
$n - 2$ square, so there are $(n - 2)^2$ of these for each of the 6 faces, so
$6(n - 2)^2$ have 1 side painted. For the cubes with 2 sides painted, these
are along each of the 12 edges, but there are $n - 2$ of these (can you see
why?), making for a total of $12(n - 2)$ of these. Finally, no matter how
large the cube is, there is only one cube with 3 sides painted per each
of the 8 corners, so there are 8 cubes with three sides painted.

Beans and Bowls

How many ways are there to arrange 10 beans among
3 bowls? Try it for different numbers of beans.

Solution:
There are lots of ways to do this problem. One way is to break it into
11 cases, based on how many beans are in the first bowl. (Can you see
why it is 11 cases and not 10 cases?) Then for each case, count how
many ways the beans can be distributed among the other 2 bowls. I
recommend doing this, so that you understand the problem more by
immersing yourself in it. Once you have done that, here s a quicker
and more elegant way to see the final answer:

Imagine your beans as dots, all lined Up, but with *2 extras:*

$$\bullet\ \bullet\ \bullet\ \bullet\ \bullet\ \bullet\ \bullet\ \bullet\ \bullet\ \bullet\ \bullet\ \bullet$$

Why 2 extras? Well, imagine that it's your job to replace 2 of the
beans with an x, like this:

$$\bullet\ \bullet\ x\ \bullet\ \bullet\ \bullet\ x\ \bullet\ \bullet\ \bullet\ \bullet$$

or this:

$$\bullet\ \bullet\ \bullet\ \bullet\ \bullet\ \bullet\ \bullet\ x\ \bullet\ x\ \bullet\ \bullet$$

or even this:

$$\bullet\ \bullet\ \bullet\ \bullet\ x\,x\ \bullet\ \bullet\ \bullet\ \bullet\ \bullet\ \bullet$$

These x's are instructions on which bowls to put the beans into. They
work as follows: put beans into the first bowl until you hit the first x;
then put beans into the second bowl until you hit the second x; then
put the rest of the beans into the third bowl. So in the first example
above, there would be 2 beans in the first bowl, 3 beans in the second
bowl, and 5 beans in the third bowl. For the second example, there

would be 7 beans in the first bowl, 1 bean in the second bowl, and 2 beans in the third bowl. For the last example, there would be 4 beans in the first bowl, 0 beans in the second bowl (do you see why?), and 6 beans in the third bowl.

So each way of replacing 2 of the 12 beans with an x corresponds to a way to put the remaining 10 beans in bowls. So if we can figure out how many ways there are to choose 2 beans and replace them with an x, then we will know how many arrangements of beans in bowls there are. How many ways are there to pick the first bean? Well, there are 12, because you can pick any bean. Then there are 11 ways to pick the second bean, because you've already picked 1. So there are $12 \times 11 = 132$ ways to pick 1 bean and then pick another bean. But, there is another thing we need to be careful of. There are 2 ways to pick each pair of x's because you can switch the order in which you picked them. So 132 counts every pair of x's twice, and we need to divide by 2. So the number of ways of replacing 2 beans by x's, which is the same as the number of ways to put 10 beans in 3 bowls, is $(12 \times 11)/2 = 66$.

Can you see how this method would work if you changed the number of beans? What about if you changed the number of bowls?

Partitions

You could use Cuisenaire rods to help with this problem.

The number 3 can be broken up into positive numbers in four different ways:

$1 + 1 + 1$

$1 + 2$

$2 + 1$

3

Or maybe you think that 1 + 2 and 2 + 1 are the same, so there are really only three ways to break up the number.

Decide which you like better and investigate partitions for different numbers using your rules.

Solution:

For this one, you're on your own! Coming up with a general pattern for how many partitions a number has is an unsolved problem. Welcome to cutting-edge mathematics.

Notes

Introduction

1. In 2006, 38.4% of A-level students in England were female, in Wales 40.8% were female, in Northern Ireland 46.2% were female and in Scotland 48.4% of Higher students were female.
2. Kounine, L., Marks, J., & Truss, E. (June, 2008). *The Value of Mathematics*: REFORM.
3. 'In 2007, 60,093 UK students took mathematics A-level compared to 84,744 in 1989. The number of A-level mathematics entries as a proportion of total A-level entries fell by two-thirds.' (Kounine, Marks & Truss, 2008, p14)
4. A government estimate, cited in Kounine, Marks & Truss, 2008, p15.
5. UNICEF. (2007). *An overview of child well-being in rich countries: A comprehensive assessment of the lives and well-being of children and adolescents in the economically advanced nations.* Florence: UNICEF Innocenti Research Centre.
6. A survey of 2,000 children carried out by Bliss magazine, reported: http://news.bbc.co.uk/cbbcnews/hi/uk/newsid_3665000/3665743.stm
7. PISA (Pisa) performance tables are based on tests taken by 15-year-olds which aim to assess their ability to apply their knowledge in "real world" situations. http://news.bbc.co.uk/1/hi/education/7115692.stm
8. A survey of 2,000 children carried out by Bliss magazine, reported: http://news.bbc.co.uk/cbbcnews/hi/uk/newsid_3665000/3665743.stm

9. Glenn, J. (2000). *Before It's Too Late*. A Report to the Nation from the National Commission on Mathematics and Science Teaching for the 21st Century. http://www.ed.gov/americacounts/glenn/

10. Kounine, L., Marks, J., & Truss, E. (June, 2008). *The Value of Mathematics*: REFORM. P15.

11. Quoted in Ball, S. (1990). Politics and Policy Making in Education. London: Routledge, p103.

12. Steen, L.A. (ed) (1997). *Why Numbers Count: Quantitative Literacy for Tomorrow's America*. New York: College Entrance Examination Board.

13. ibid

14. Moses, R., & Cobb, J. C. (2001). *Radical Equations: Math, Literacy and Civil Rights*. Boston: Beacon Press.

15. Steen, L.A. (ed) (1997). *Why Numbers Count: Quantitative Literacy for Tomorrow's America*. New York: College Entrance Examination Board.

16. Gainsburg, J. (2003). The Mathematical Behavior of Structural Engineers. Stanford University, Stanford, California. Dissertation Abstracts International, A 64/05, p34.

17. Hoyles, C., Noss, R., & Pozzi, S. (2001). Proportional Reasoning in Nursing Practice. *Journal for Research in Mathematics Education, 32*(1), 4-27.

18. Nunes, T., Schliemann, A. D., & Carraher, D. W. (1993). *Street Mathematics and School Mathematics*. New York: Cambridge University Press.

19. Gainsburg, J. (2003). The Mathematical Behavior of Structural Engineers. Stanford University, Stanford, California. Dissertation Abstracts International, A 64/05, p36.

20. Hoyles, C., Noss, R., & Pozzi, S. (2001). Proportional Reasoning in Nursing Practice. *Journal for Research in Mathematics Education, 32*(1), 4-27.

21. Lave, J., Murtaugh, M. & de la Rocha, O. (1984). The Dialectical Construction of Arithmetic Practice. In B. Rogoff & J. Lave. (eds)., *Everyday Cognition: It's Development in Social Context*. Cambridge, Mass.: Harvard University Press.

22. Lave, J. (1988). *Cognition in practice*. Cambridge: Cambridge University Press.

23. Masingila, J. (1993). Learning from Mathematics Practice in Out-of-School Situations. *For the Learning of Mathematics, 13*(2), 18-22.

24. Nunes, T., Schliemann, A. D., & Carraher, D. W. (1993). *Street Mathematics and School Mathematics*. New York: Cambridge University Press.

25. Lave, J. (1988). *Cognition in practice*. Cambridge: Cambridge University Press.

26. Boaler, J. (2002). *Experiencing School Mathematics: Traditional and Reform Approaches to Teaching and Their Impact on Student Learning.* (Revised and Expanded Edition ed.). Mahwah, NJ: Lawrence Erlbaum Association.

1/ What is Maths? And Why Do We *All* Need It?

1. Brown, Dan. (2003) *The Divinci Code.* New York: Doubleday.
2. Wertheim, M. (1997) *Pythagoras' Trousers: God, Physics and the Gender Wars.* New York, W.W. Norton & Company.pp.3-4
3. Boaler, J. (2002). *Experiencing School Mathematics: Traditional and Reform Approaches to Teaching and Their Impact on Student Learning.* (Revised and Expanded Edition ed.). Mahwah, NJ: Lawrence Erlbaum Association.
4. Devlin, K. (2000) *The Math Gene: How Mathematical Thinking Evolved and Why Numbers are Like Gossip* (Basic Books: New York).p.7.
5. Kenschaft, Patricia Clark (2005). *Math Power: How to Help Your Child Love Math, Even if You Don't.* Revised Edition. Upper Saddle River, NJ, Pi Press.
6. Sawyer, W.W. (1955) *Prelude to Mathematics.* New York: Dover Publications.p.12.
7. Fiori, N. (2007). *The practices of mathematicians.* Manuscript in preparation.
8. Singh, S. (1997) *Fermat's Enigma: The Epic Quest to solve the world's greatest mathematical problem.* New York: Anchor Books.
9. Ibid.
10. Singh, S. (1997) *Fermat's Enigma: The Epic Quest to solve the world's greatest mathematical problem.* New York: Anchor Books. P. xiii
11. Quoted in Singh, S. (1997) *Fermat's Enigma: The Epic Quest to solve the world's greatest mathematical problem.* New York: Anchor Books. P. 6
12. Article in New York Times by Fran Schumer called *In Princeton, Taking On Harvard's Fuss About Women.* New York Times, June 19, 2005. In it she quotes Diane Maclagan.
13. Lakatos, I. (1976) *Proofs and refutations.* Cambridge, UK, Cambridge University Press.
14. Cockcroft, W. H. (1982). *Mathematics Counts: Report of Inquiry into the Teaching of Mathematics in Schools.* London: HMSO.
15. Albers, D. J., Alexanderson, G. L., & Reid, C. (1990). *More mathematical people: contemporary conversations.* (Boston, Harcourt Brace Jovanovich),p.30
16. Devlin, K. (2000) *The Math Gene: How Mathematical Thinking Evolved and Why Numbers are Like Gossip* (Basic Books: New York).p.76.
17. Pólya, G (1971). *How to solve it.* New York: Doubleday Anchor.,v.
18. Burton, L. (1999) The Practices of Mathematicians: What do they tell us about coming to know mathematics? *Educational Studies in Mathematics,* 37, p36.

19. *Peter Hilton, popular quote.*
20. Hersh, R. (1997) *What is Mathematics, Really?* New York, Oxford University Press.p.18.
21. Devlin, K. (2000) *The Math Gene: How Mathematical Thinking Evolved and Why Numbers are Like Gossip* (Basic Books: New York).p.9.
22. Fiori, N. (2007). *The practices of mathematicians.* Manuscript in preparation.
23. Boaler, J. (2002). *Experiencing School Mathematics: Traditional and Reform Approaches to Teaching and Their Impact on Student Learning.* (Revised and Expanded Edition ed.). Mahwah, NJ: Lawrence Erlbaum Association.
24. Reid, D.A. (2002) Conjectures and Refutations in Grade 5 Mathematics. *Journal for Research in Mathematics Education.* 33(1), 5-29.
25. Keil, G.E. (1965) *Writing and solving original problems as a means of improving verbal arithmetic problem solving ability.* Unpublished doctoral dissertation. Indiana University

2 / What's Going Wrong in Classrooms? Identifying the Problems

1. Dixon, A. (2002) Editorial, *FORUM, 44(1), p.1.*
2. PISA (Pisa) performance tables are based on tests taken by 15-year-olds which aim to assess their ability to apply their knowledge in "real world" situations. http://news.bbc.co.uk/1/hi/education/7115692.stm
3. Brown, M. (2008). *Full Report of Research Activities and Results, Students' Experiences of Undergraduate Mathematics.* Reference Number: R000238.
4. UNICEF. (2007). *An overview of child well-being in rich countries: A comprehensive assessment of the lives and well-being of children and adolescents in the economically advanced nations.* Florence: UNICEF Innocenti Research Centre.
5. Boaler, J. (1997). *Experiencing School Mathematics: Teaching Styles, Sex and Setting.* Buckingham: Open University Press.
6. Carpenter, T.P., Franke, M.L., & Levi, L. (2003). *Thinking mathematically: Integrating arithmetic and algebra in the elementary school.* Portsmouth, NH: Heinemann.
7. Flannery, S. (2002) *In Code: A Mathematical Journey.* Chapel Hill: Algonquin Books. p38.
8. Hersh, R. (1997) *What is Mathematics, Really?* New York: Oxford University Press.p.27.
9. Boaler, J. (2008). *What's Math Got To Do With It? Helping Children Learn to Love Their Least Favorite Subject – and Why It's Important for America.* Viking: New York.
10. Boaler, J., & Greeno, J. (2000). Identity, Agency and Knowing in Mathematics Worlds. In J. Boaler (Ed.), *Multiple Perspectives on Mathematics Teaching and Learning* (pp. 171-200). Westport:CT: Ablex Publishing.
11. Murata, Aki 2006. Personal Communication.

12. Schoenfeld, A. H. (1987). Confessions of an accidental theorist. *For the Learning of Mathematics,* 7(1), 30-38.p.37.
13. Rose, H. (1998) Reflections on PUS, PUM and the weakening of Panglossian cultural tendencies. *The Production of a Public Understanding of Mathematics.* Birmingham: University of Birmingham.p.4
14. Ibid.

3 / A Vision for a Bettter Future: Effective Classroom Approaches

1. The names of all schools, teachers and students involved in the research studies reported in this book are pseudonyms.
2. Siskin, L., S. (1994). *Realms of Knowledge: Academic departments in Secondary Schools.* London, Falmer Press.
3. Wenger, E. (1998). *Communities of Practice: Learning, Meaning and Identity.* Cambridge, Cambridge University Press.
4. Lave, J. (1988*). Cognition in practice.* Cambridge, Cambridge University Press.
5. Boaler (2005). The 'Psychological Prison' from which they never escaped: The role of ability grouping in reproducing social class inequalities. *FORUM, 47, 2&3, 135-144.*
6. Boaler, J & Staples, M. (2008). Creating Mathematical Futures through an Equitable Teaching Approach: The Case of Railside School. *Teachers' College Record.* 110 (3), 608-645.

4/ Banishing the Monsters: Moving to More Effective Forms of Assessment

1. These are the standardized tests given to children at age 11 in England.
2. Preceded by the O-level and CSE exams
3. Mansell, W. (2007). *Education By Numbers: The Tyranny of Testing.* London: Politico's Publishing.
4. Mansell, W. (2007). *Education By Numbers: The Tyranny of Testing.* London: Politico's Publishing.
5. Reay, D., & Wiliam, D. (1999). I'll be a nothing: structure, agency and the construction of identity through assessment. *British Educational Research Journal, 25*(3), p345-346.
6. Mansell, W. (2007). *Education By Numbers: The Tyranny of Testing.* London: Politico's Publishing.
7. Mansell, W. (2007). *Education By Numbers: The Tyranny of Testing.* London: Politico's Publishing. p25
8. Amrein, A.L., & and Berliner, D. (2002). High-Stakes Testing: Uncertainty and Student Learning. *Education Policy Analysis Archives* 10(18), pp 1-63.
9. The researchers used four different measures: the ACT administered by

the American College Testing Program, the SAT administered by the College Board, the NAEP (National Assessment of Educational Progress), administered by the US Department of Education and the AP (Advanced Placement) exams, administered by the College Board.

10. *A Thousand Words to Shape the Future.* A Response to QCA from The Mathematical Association, p1. (http://www.m-a.org.uk/whats_new/A-thousand-words/a-thousand-words.htm).

11. *Making mathematics count: the report of Sir Adrian Smith's inquiry into post-14 mathematics education* (937764), The Stationery Office, February 2004.

12. OFSTED. *Evaluating mathematics provision for 14–19-year-olds.* Crown Copyright, 2006. Ref Number: HMI 2611.

13. Kohn, A. (2000). *The Case against Standardized Testing.* Portsmouth, NH:Heinemann. P30.

14. The Mathematical Association and the Association for Teachers of Mathematics.

15. For information about Scotland's implementation of 'assessment for learning' see http://www.ltscotland.org.uk/assess/about/index.asp

16. Black, P., & Wiliam, D. (1998). Inside the Black Box: Raising Standards through Classroom Assessment. *Phi Delta Kappan, October,* 139-148.

17. Black, P., & Wiliam, D. (2005). Lessons from around the world: how policies, politics and cultures constrain and afford assessment practices *Curriculum Journal, 16*(2), 249-261.

18. White, B. Y., & Frederiksen, J. R. (1998). Inquiry, Modeling, and Metacognition: Making Science Accessible to All Students. *Cognition and Instruction, 16*(1), 3-118.

19. Advanced Placement (AP) Physics tests – with content at a similar level to A-level physics.

20. Black, P., Harrison, C., Lee, C., Marshall, B., & Wiliam, D. (2002). *Working inside the black box: assessment for learning in the classroom.* London: Dept of Education & Professional Studies, King's College.

21. Elawar, M.C., & Corno, L. (1985). A Factorial Experiment in Teachers' Written Feedback on Student Homework: Changing Teacher Behavior a Little Rather Than a Lot. *Journal of Educational Psychology,* 77 (2), pp162-173.

22. Butler, R. (1988). Enhancing and Undermining Intrinsic Motivation: The Effects of Task-Involving and Ego-Involving Education on Interest and Performance. *British Journal of Educational Psychology,* 58, pp1-14.

23. Wiliam, D. (2007). Keeping Learning on Track: Classroom Assessment and the Regulation of Learning. In K.F. Lester Jr (ed). *Second Handbook*

of Mathematics Teaching and Learning (pp1053-1098). Greenwich, Conn,: Information Age Publishing, p. 1085.

24. Sadler, R. (1989). Formative Assessment and the Design of Instructional Systems. *Instructional Science,* 18, pp 119-144.

25. Hutchinson, Carolyn and Hayward, Louise (2005) The journey so far: assessment for learning in Scotland', *Curriculum Journal,* 16:2, 225-248

26. Black, Paul (2005) 'Formative assessment: views through different lenses', *Curriculum Journal,* 16:2, p134.

27. For example, The Association of Teachers of Mathematics (ATM), The Mathematical Association (MA), The Institute of Mathematics and its Applications (IMA), and the London Mathematical Society (LMS).

5 / Making 'Low Ability' Children: How Different Forms of Grouping Can Make or Break Children

1. Devlin, K. (2001). *The Maths Gene: Why everyone has it, but most people don't use it.* Phoenix: London.

2. Blatchford, P., Hallam, S., Ireson, J., Kutnick, P., & Creech, A. (2008). Classes, Groups and Transitions: Structures for Teaching and Learning. Research Survey 9/2 *The Primary Review.* Cambridge: University of Cambridge.

3. Kutnick, P., Sebba, J., Blatchford, P., Galton, M., & Thorp. J. (2007). *The Effects of Pupil Grouping: Literature Review.* Department for Education and Skills: London.

4. Blatchford, P., Hallam, S., Ireson, J., Kutnick, P., & Creech, A. (2008). Classes, Groups and Transitions: Structures for Teaching and Learning. Research Survey 9/2 *The Primary Review.* Cambridge: University of Cambridge.

5. The Primary Review. http://www.primaryreview.org.uk

6. Blatchford, P., Hallam, S., Ireson, J., Kutnick, P., & Creech, A. (2008). Classes, Groups and Transitions: Structures for Teaching and Learning. Research Survey 9/2 *The Primary Review.* Cambridge: University of Cambridge. p28.

7. Ibid, p28.

8. UNICEF. (2007). *An overview of child well-being in rich countries: A comprehensive assessment of the lives and well-being of children and adolescents in the economically advanced nations.* Florence: UNICEF Innocenti Research Centre.

9. Beaton, A.E. & O'Dwyer, L.M. (2002) Separating School, Classroom, and Student Variances and their Relationship to Socio-economic Status, in: D.F. Robitaille & A.E. Beaton (Eds) *Secondary Analysis of the TIMSS Data* (Dordrecht, Kluwer Academic Publishers).

10. Yiu, L. (2001) *Teaching Goals of Eight Grade Mathematics Teachers: Case*

Study of Two Japanese Public Schools. School of Education (Stanford, Stanford University).

11. Burris, C., Heubert, J. & Levin, H. (2006) *American Educational Research Journal, 43(1), pp. 103-134.*

12. Porter, A.C. and associates. (1994) *Reform of high school mathematics and science and opportunity to learn.* (New Brunswick, Consortium for Policy Research in Education).

13. Rosenthal, R. & Jacobson, L. (1968) *Pygmalion in the classroom* (New York, Holt, Rinehart & Winston).

14. Boaler, J., Wiliam, D. & Brown, M. (2000) Students' Experiences of Ability Grouping-disaffection, polarisation and the construction of failure, *British Educational Research Journal*, 26(5), pp. 631-648.

15. PISA, 2003 *Learning from Tomorrow's World: First Results from PISA 2003.* PISA:OECD

16. Boaler, J & Staples, M. (2008). Creating Mathematical Futures through an Equitable Teaching Approach: The Case of Railside School. *Teachers' College Record.* 110 (3), 608-645; Oakes, J. (2005). *Keeping Track. How Schools Structure Inequality.* (Second Edition ed.). New Haven: Yale University Press.

17. Wiliam, D. & Bartholomew, H. (2004) It's not which school but which set you're in that matters: the influence of ability grouping practices on student progress in mathematics, *British Educational Research Journal*, 30(2), pp. 279-293.

18. Dunne, M., Humphreys, S., Sebba, J., Dyson, A., Gallannaugh, F., & Muijs, D. *Effective Teaching and Learning for Pupils in Low Attaining Groups.* Research Report DCSF-RR011 London: Department for Children, Schools and Families.

19. Boaler (2005). The 'Psychological Prison' from which they never escaped: The role of ability grouping in reproducing social class inequalities. *FORUM*, 47, 2&3, 135-144.

20. Oakes, J. (2005) *Keeping Track. How Schools Structure Inequality.* (New Haven, Yale University Press).p.217.

21. Ibid.p.218.

22. Olson, S. (2005) *Countdown: the race for beautiful solutions at the International Mathematical Olympiad* (New York, Houghton Mifflin).pp.48-49.

23. Boaler, J. (2006) The 'Psychological Prison' from which they never escaped: The role of ability grouping in reproducing social class inequalities, *FORUM*, 47(2&3), pp. 135-144.

24. Dixon, A. (2002) Editorial, *FORUM*, 44(1), p.1.

6 / Paying the Price for Sugar and Spice? How Girls and Women Are Kept Out of Maths and Science

1. Chi-squared = 16.96, n=163, 4 d.f. p < 0.001
2. Zohar, A., & Sela, D. (2003). Her physics, his physics: gender issues in Israeli advanced placement physics classes. *International Journal of Science Education, 25*(2),
3. Ibid.
4. Gilligan, C. (1982). *In a Different Voice: Psychological Theory and Women's Development*. Cambridge, Massachusetts: Harvard University Press.
5. Belenky, M.F., Clinchy, B.M., Golderberger, N.R, & Tarule, J.M. (1986). *Women's Ways of Knowing*. NY: Basic Books, 214-229.
6. Brizendine, L. *The Female Brain*. New York: Morgan Road Books., p.4.
7. Ibid.
8. Ibid.
9. Ibid.
10. Connellan, J. & Baron-Cohen, S. (2000). Sex Differences in Human Neonatal Social Perception. *Infant Behavior & Development*, 23, 113-118, p114.
11. Reported in Brizendine, L. *The Female Brain*. New York: Morgan Road Books, p.127
12. Sax, L. *Why Gender Matters: What parents and teachers need to know about the emerging sciences of sex differences*. New York: Broadway Books.p.31.
13. Ibid., p.101.
14. Boaler, J. & Kent, G. (in press). Mathematics and Science in the United Kingdom: Inequities in Participation and Performance. *The Royal Society*.
15. Hyde, J. S., Fennema, E., & Lamon, S. (1990). Gender Differences in Mathematics Performance: A Meta-Analysis. *Psychological Bulletin, 107*(2), 139-155.
16. They recorded an effect size of only +0.15 standard deviations.
17. Becker, J. (1981). Differential Treatment of Females and Males in Mathematics Class. *Journal for Research in Mathematics Education, 12*(1), 40-53.
18. Herzig, A. (2004a). Becoming mathematicians: Women and students of color choosing and leaving doctoral mathematics. *Review of Educational Research,, 74*(2), 171-214.
19. Herzig, A. (2004b). Slaughtering this beautiful math: graduate women choosing and leaving mathematics. *Gender and Education, 16*(3), 379-395.
20. Cohen, M. (1999). 'A habit of healthy idleness': boys' underachievement in historical perspective. In D.Epstein, J.Elwood & V.Hey (Eds.), *Failing Boys? Issues in Gender and Achievement*. Buckingham, England: Open University Press.p.24.

21. Ibid.,p.25.

22. Rogers, P., & Kaiser, G. (Eds.). (1995). *Equity in Mathematics Education: Influences of Feminism and Culture*. London: Falmer Press.

7 / Talking with Numbers: Key Strategies and Ways of Working

1. Gray, E., & Tall, D. (1994). Duality, Ambiguity, and Flexibility: A "Proceptual" View of Simple Arithmetic. *Journal for Research in Mathematics Education, 25*(2), pp.116-140.

2. Thurston, W.P. (1990). Mathematical Education. *Notices of the American Mathematical Society*, 37, 844-850.

3. Beck, T. A. (1998). Are there any questions? One teacher's view of students and their questions in a fourth-grade classroom. *Teaching and Teacher education, 14*(8), pp.871-886

4. Good, T. L., Slavings, R. L., Harel, K. H., & Emerson, H. (1987). Student Passivity: A Study of Question Asking in K-12 Classrooms. *Sociology of Education, 60*(July), pp.181-199.

5. Boaler, J., & Staples, M. (2008). Creating Mathematical Futures through an Equitable Teaching Approach: The Case of Railside School. *Teachers' College Record, 110* (3), 608-645.

6. Greeno, J. G. (1991). Number sense as situated knowing in a conceptual domain. *Journal for Research in Mathematics Education, 22*(3), pp.170-218.

7. Boaler, J., & Humphreys, C. (2005). *Connecting Mathematical ideas: Middle school video cases to support teaching and learning*. Portsmouth, NH: Heinemann.

8. Ball, D. L. (1993). With an Eye on the Mathematical Horizon: Dilemmas of Teaching Elementary Mathematics. *The Elementary School Journal, 93*(4), pp.373-397.

9. Lampert, M. (2001). *Teaching problems and the problems of teaching*. New Haven: Yale University Press

10. Kent, P., & Noss, R. (2000). The visibility of models: using technology as a bridge between mathematics and engineering. *International Journal of Mathematics Education, Science & Technology., 31*(1), pp.61-69.

11. Noss, R., & Hoyles, C. (2002). Abstraction in Expertise: A Study of Nurses' Conceptions of Concentration. *Journal for Research in Mathematics Education, 33*(3), pp.204-229.

12. I am grateful to Emily Shahan, for her in-depth analysis and descriptions of Jorge's experiences.

13. I am grateful to Tesha Sengupta-Irving, for her in-depth analysis and descriptions of Alonzo's experiences.

8 / Giving Children the Best Mathematical Start in Life: Activities and Advice

1. Albers, D. J., Alexanderson, G. L., & Reid, C. (Eds.), (1990). *More Mathematical People: Contemporary Conversations* Boston: Harcourt Brace Jovanovich.
2. Fiori, N. (2007). *In Search of Meaningful Mathematics: The Role of Aesthetic Choice.* Doctoral Dissertation, Stanford University.
3. Casey, M. B., Nuttall, R. L., & Pezaris, E. (1997). Mediators of gender differences in mathematics college entrance test scores: A comparison of spatial skills with internalized beliefs and anxieties. *Developmental Psychology, 33*, 669 – 680.
4. Duckworth, E., (1996). *"The having of wonderful ideas" and other essays on teaching and learning.* New York: Teacher's College Press.
5. Ryan, A. M., & Patrick, H. (2001). The Classroom Social Environment and Changes in Adolescents' Motivation and Engagement During Middle School. *American Educational Research Journal, 38*(2), pp.437-460.
6. Eccles, J., Wigfield, A., Midgley, C., Reuman, D. A., Mac Iver, D., & Feldlaufer, H. (1993). Negative Effects of Traditional Middle Schools on Students' Motivation. *The Elementary School Journal, 93*(5), pp.553-574.
7. Stipek, D. & Seal, K. (2001). *Motivated Minds. Raising Children to Love Learning.* New York: Henry Holt and Company.
8. Frank, M. (1988). Problem solving and mathematical beliefs. *Arithmetic Teacher, 35,* 32-34.
9. Garofalo, J. (1989). Beliefs and their influence on mathematical performances. *Mathematics Teacher, 82* (7), 502-505.
10. Flannery, S. (2002). *In Code: A Mathematical Journey.* Chapel Hill: Algonquin Books.
11. Ibid. p.8.
12. Ibid.
13. Kenschaft, P.C. (2006). *Math Power. How to help your child love math, even if you don't.* New York: Pi Press. p50.
14. Boaler, J., & Humphreys, C. (2005). *Connecting Mathematical ideas: Middle school video cases to support teaching and learning.* Portsmouth, NH: Heinemann.
15. Kenschaft, P.C. (2006). *Math Power. How to help your child love math, even if you don't.* New York: Pi Press. P51.
16. Eccles, J., & Jacobs, J. E. (1986). Social forces shape math attitudes and performance. Signs. *Journal of Women in Culture and Society, 11*(21), pp.367-380.
17. Kenschaft, P.C. (2006). *Math Power. How to help your child love math, even if you don't.* New York: Pi Press. p35.

18 Beilock, S. L., Holt, L. E., Kulp, C. A., & H., C. T. (2004). More on the Fragility of Performance: Choking Under Pressure in Mathematical Problem Solving. *Journal of Experimental Psychology, 133*(4), pp.584-600.

19. Pólya, G (1957). *How to solve it*. New York: Doubleday Anchor.

20. Lee, K.S. (1982). Fourth Graders' Heuristic Problem-Solving Behavior. *Journal for Research in Mathematics Education,* 13 (2), pp. 110-123.

21. RAND, M. S. P. (2002, October). *Mathematical proficiency for all students: Toward a strategic research and development program in mathematics education (DRU-2773-OERI)*. Arlington, VA: RAND Education & Science and Technology Policy Institute, px.

22. Lee, K.S. (1982). Fourth Graders' Heuristic Problem-Solving Behavior. *Journal for Research in Mathematics Education,* 13 (2), pp. 110-123.

23. Gray, G., & Tall, D. (1994). Duality, Ambiguity and Flexibility: A "Proceptual" View of Simple Arithmetic. *Journal for Resaerch in Mathematics Education,* 25(2), 99, 1160140.

9 / Making a Difference through Work with Schools

1. Maher, C. (1991). Is Dealing with Mathematics as a Thoughtful Subject Compatible with Maintaining Satisfactory Test Scores? A Nine-Year Study. *Journal of Mathematical Behaviour, 10*, pp. 225-248; Boaler, J. (1997). *Experiencing School Mathematics: Teaching Styles, Sex and Setting*. Buckingham: Open University Press.

2. Kenschaft, P.C. (2006). *Math Power. How to help your child love math, even if you don't.* New York: Pi Press.

3. Kenschaft, P.C. (2006). *Math Power. How to help your child love math, even if you don't.* New York: Pi Press.

4. Maher, C. (1991). Is Dealing with Mathematics as a Thoughtful Subject Compatible with Maintaining Satisfactory Test Scores? A Nine-Year Study. *Journal of Mathematical Behaviour, 10*, pp. 225-248; Boaler, J. (2002). *Experiencing School Mathematics: Traditional and Reform Approaches to Teaching and Their Impact on Student Learning.* (Revised and Expanded Edition ed.). Mahwah, NJ: Lawrence Erlbaum Association.; Riordan, J. E., & Noyce, P. E. (2001). The impact of two standards-based mathematics curricula on student achievement in Massachusetts. *Journal for Research in Mathematics Education, 32*(4), pp.368-398.; Schoenfeld, A. H. (2002, January/February). Making mathematics work for all children: Issues of standards, testing, and equity. *Educational Researcher,* 31(1), 13-25.

5. http://www.i-nfer.co.uk/package.html

6. http://balancedassessment.concord.org/

Index